The Best American Science and Nature Writing 2017

The Best American Science and Nature Writing™ 2017

Edited and with an Introduction
by Hope Jahren

Tim Folger, *Series Editor*

A Mariner Original
HOUGHTON MIFFLIN HARCOURT
BOSTON • NEW YORK 2017

hmhco.com

ISBN 978-1-328-71551-7 (print)   ISSN 1530-1508 (print)
ISBN 978-1-328-71556-2 (ebook)   ISSN 2573-475X (ebook)

Printed in the United States of America
DOC 10 9 8 7 6 5 4 3 2 1

Contents

PART III. The "Real Life" of Scientists

Foreword

MODERN COSMOLOGY WAS born in Germany a century ago, and within two decades of its birth it almost died there. When Albert Einstein published his general theory of relativity in November 1915, it's doubtful he could have imagined how profoundly deranged his country would become. On May 10, 1933—the same year Einstein left Germany forever—mobs of young Nazis and their supporters across Germany were feeding bonfires with his papers, along with works by Sigmund Freud, Thomas Mann, Bertolt Brecht, Erich Maria Remarque, and others supposedly contaminated with *undeutschen Geist*—un-German spirit. More than 25,000 books burned on that day, including those of the 19th-century Jewish poet and playwright Heinrich Heine, who had once written, "Where they burn books, they will also ultimately burn people."

Einstein's own research—which transformed our understanding of the universe—was condemned by vicious ideologues as an example of "Jewish physics," whatever that was supposed to be. Even Werner Heisenberg, one of the founders of quantum theory, could not summon the courage to defend Einstein. He capitulated to Nazi authorities, going so far as to request that his colleagues not mention the great physicist's name in their lectures or publications. Heisenberg himself practiced that same self-censorship in all his work and talks while the Nazis remained in power.

Not all German scientists followed Heisenberg's spineless lead. Max von Laue, a Nobel laureate described by one colleague as "a sensitive and even a nervous man," nevertheless openly opposed Nazi policies. He publicly compared the attacks on Einstein to the Inquisition's censure of Galileo. Some accounts say that von Laue always carried something while out walking so he could avoid the mandatory "Heil Hitler" salute. He even helped some colleagues escape from Germany. The Nazis eventually forced von Laue to resign from his university position, in 1943, but he remained in Germany during and after the war, where he helped rebuild a shattered scientific community, a community partly undone by its own leaders.

One measure of the health of any modern society must be the degree to which it supports its scientists. A few days before I started to write this foreword, hundreds of thousands of people in dozens of cities across the country participated in the March for Science. It was an event at once inspiring and worrisome: inspiring because so many took a stand for rationalism—a public rebuke to the nation's leaders that couldn't be more different from the German book burnings of the 1930s; worrisome because who would have thought that in the 21st century scientists and citizens would feel the need to gather in support of something so self-evidently valuable as unfettered scientific research?

Yet the march was necessary, urgently so. Scientists at more than a dozen federal agencies have launched rogue Twitter feeds to counter the policies of a frighteningly uninformed president who once tweeted that "global warming was created by and for the Chinese." We live at a pivotal moment in history, and not just for ourselves. We have already pumped enough greenhouse gases into the atmosphere that the planet will continue to heat up for centuries to come. Our inaction has fated future generations to a world with flooded coasts, extreme weather, and other catastrophes that will trigger political, social, and economic instability. Scientists are a cautious lot, but some of them now warn that climate change threatens not just "the environment" but civilization itself.

Tragically, we're now led—if that's the word—by a government that denies the very existence of the crisis that is upon us. The president has promised to scrap funding for NASA's vitally important climate-monitoring satellites. Scott Pruitt, the new head of the Environmental Protection Agency, told reporters earlier this year, "I do not agree that it's [carbon dioxide] a primary contributor to the global warming that we see." Given his position, and his denial of basic scientific facts, Pruitt may well be one of the most dangerous people on the planet. He would do well to heed one of Philip K. Dick's aphorisms: "Reality is that which, when you stop believing in it, doesn't go away." Without the active participation of the United States in global climate agreements over the next few years, we risk losing forever any chance of avoiding the most disruptive effects of climate change. Our descendants will not judge us kindly.

Sustained, in-depth reporting on these issues has never been more important. And few journalists have done as much as Elizabeth Kolbert to highlight the enormity of the threat posed by global warming. Her contribution to this collection, "A Song of Ice," recounts her trip to Greenland's vast ice sheet, which covers about 80 percent of the country, with a height exceeding 10,000 feet in the interior. "The ice sheet is so big," she writes, "that it creates its own weather. Its mass is so great that it deforms the earth, pushing the bedrock several thousand feet into the mantle. Its gravitational tug affects the distribution of the oceans." Among the scientists she met was a group studying a single melt stream in northeastern Greenland. That one melt stream, the scientists told her, will eventually raise sea levels around the world by more than three feet, high enough so that harbor waves would lap at the base of Manhattan's new World Trade Center.

Tom Kizzia takes us on another Arctic journey in his remarkable story, "The Last Harpoon," about Inupiat whale hunters in northern Alaska. Their community of "small but comfortable homes, laid out around a new school and a diesel-fired power plant," depends for its survival on fossil fuels, the very substances that threaten to destroy their ancient hunting traditions.

In "The Invisible Catastrophe," Nathaniel Rich writes about a methane leak at a natural gas facility in Southern California. Methane, Rich tells us, is a potent greenhouse gas, with a warming effect more than 80 times that of carbon dioxide. By the time state officials announced that the leak had finally been capped early in 2016, it had already released enough methane to equal the global-warming impact of the exhaust belched by nearly 2 million cars over the course of a year.

The articles mentioned above remind us—as do all the stories in this collection—of something easily overlooked: science is an intensely personal pursuit. All the data and discoveries, all the remarkable insights about the world and our place in it, come from people beset with the same worries, ambitions, and career obstacles familiar to each of us. And scientists are not immune to the ills that plague any other profession. Hope Jahren, our guest editor, has included two unsettling stories that focus on an issue that has long been underreported: sexual harassment in science. As Azeen Ghorayshi and Kathryn Joyce show in their forceful articles, scientists sometimes need protection not just from malevolent government officials but from their own predatory colleagues and mentors.

Hope has organized her selections in this anthology under three broad themes: Emergent Fields; Changing Land and Resources; and The "Real Life" of Scientists. You'll find in the pages ahead accounts of an astronomer's search for worlds like our own, the ongoing struggle to save endangered species, the tragic story of a scientist obsessed with altruism, and more than a dozen other examples of some of the best and most important writing of our time.

As we finish work on this year's anthology, I can't help wondering what this year will bring. Will the United States, against all expectations, take the lead in confronting climate change? Maybe next year's collection will include stories that will surprise us all—in a good way. To that end, I'm already gathering candidates for the 2018 edition. And you, readers, can help! I promise to read widely, but I depend on many thoughtful readers, writers, and editors from around the world. Nearly every week my physical mailbox or my email inbox contains some new surprise, a story I would never have found on my own. So please, nominate your favorites for next year's anthology at http://timfolger.net/forums. I encourage writers to submit their own stories. The criteria for submissions and deadlines, and the address to which entries should be sent, can be found in the "news and announcements" forum on my website. Once again this year I'm offering an incentive to enlist readers to scour the nation in search of good science and nature writing: Send me an article that I haven't found, and if the article makes it into the anthology, I'll mail you a free copy of next year's edition. Maybe next year's guest editor will sign it for you. I also encourage readers to use the forums to leave feedback about the new collection and to discuss all things scientific. The best way for publications to guarantee that their articles are considered for inclusion in the anthology is to place me on their subscription list, using the address posted in the news and announcements forum.

It has been a privilege to work with Hope Jahren. Her first book, *Lab Girl,* published in 2016, won a tremendous number of accolades, including the National Book Critics Circle Award for best autobiography. The *New York Times* said that *Lab Girl* "does for botany what Oliver Sacks's essays did for neurology, what Stephen Jay Gould's writings did for paleontology." Not bad for a first effort! As in years past, I'm very grateful to Naomi Gibbs and her colleagues at Houghton Mifflin Harcourt, who are responsible for the entire series of Best American anthologies. And firstly and lastly, I'm grateful to my beauteous wife, Anne Nolan.

But enough from me. Please turn the page, where Hope will tell you more about why she selected these fascinating stories, which deserve the widest possible audience.

TIM FOLGER

Introduction

IN 1982 YOU could pull off of North Carolina State Highway 751, walk into a gas station, buy a pack of cigarettes and enough gas to get you to Tennessee, pay with a five-dollar bill, and still get change back. Once you got back on the road, you'd find yourself driving through what looked like the apocalypse, for 1982 was also the year that a group of men systematically severed at the base every one of the 60-foot-tall loblolly pine (*Pinus taeda*) trees planted in the area, leaving no fewer than 80 acres of pure demolition behind them. Cleanup crews mowed through the debris using huge drum choppers and then burned what was left. Each of the trees that was killed was, roughly figured, 30 to 40 years older than the man who cut it down. These fallen trees had been planted by other men in 1922, to start a forest and effectively end 60 years of cotton and tobacco sharecropping at the location. Exactly 60 years later, their little forest met its end.

But sometimes the difference between an ending and a beginning is blurry. In 1983 researchers at Duke University started yet another forest at the site by planting hundreds upon hundreds of loblolly pine seedlings within the very ashes of their predecessors. The immediate goal was homogeneity: each of the seedlings was exactly three years old, and they all were closely related genetically —the equivalent of half-siblings in human terms. The seedlings were spaced exactly two and a half meters apart. Ten years later, in 1994, some of the same researchers, plus a new generation of students, began building the most grand-scale and magnificent forest experiment that had ever been attempted. They built huge ringed scaffolds 100 feet across, taller than the forest could ever grow, and they pumped carbon dioxide from massive tankers up through the pipes that lined the scaffolds, bleeding an extra dose of carbon into the air that the little trees used for growth.

The trees loved it: carbon dioxide fueled growth, and more and more carbon meant more and more growth. The deciduous understory decorated itself gaudily with lots of cheaply made leaves, while poison ivy vented its spleen by becoming even more poisonous. Across all this diversity, however, a general trend emerged: on average, most plants grew about 30 to 40 percent bigger at the higher levels of carbon dioxide than they did when grown at the normal ambient levels. It was perhaps the most important plant experiment of the 20th century, and it has since inspired thousands of spinoff experiments, such as the ones that I do in my lab, pushing to higher and higher levels of carbon dioxide, trying to find the point when *more than enough* becomes *who cares how much*.

Duke University's Free Air CO_2 Enrichment (FACE) experiment ran for 15 years and then, rather suddenly, it was ended, as were several other open-air CO_2 experiments: the aspen forest in Wisconsin, the sweetgum plantation in Tennessee, the desert scrub in Nevada—all shut down within a couple of years. The reason that agencies cited was that the experiments were too expensive: it cost taxpayers more than $2 million each year to keep carbon dioxide moving through the scaffolding of Duke's FACE forest. But there was another, more practical reason to end the experiment that was rarely mentioned: the trees had simply outgrown their scaffolding. They had thrived so much better and faster than anyone expected that they naturally had grown across the boundaries of the test plots and out into the real world. Fortified by health and maturity and untroubled by the constraints of the past, they confidently reached out into something new.

I am sometimes charged with allegorizing human endeavors by way of plant biology, so often lately that I've decided to start actually doing it. Right here, right now, I'll suggest that the Internet is like carbon dioxide for science writers, who are themselves like plants, and that we are living in an unprecedented era of diverse and thriving journalism in the service of science—albeit one that could be cut off and mowed down if we don't actively value and defend it. My goal with this volume of *The Best American Science and Nature Writing* was to bring forward the new and unusual topics and voices of 2016, and in so doing I have focused on three main themes: Emergent Fields, Changing Land and Resources, and The "Real Life" of Scientists.

The most precious currency of science is new ideas, and so I wanted to highlight especially the journalists who brought out the newest of the new. Entire scientific fields emerged during 2016, seemingly out of the ether, and they were as disparate as they were fascinating: Maria Konnikova wrote about neurogastronomy, a biosensual new approach to making food taste good, while Nicola Twilley described how the discovery of gravitational waves has revised our understanding of spacetime forever; these two stories perhaps form the perfect contrast between the tangible and intangible delights to be had from fcience. There is also the new science that we've only recently realized we need, such as Sarah Everts's story on the new chemistry of artifact preservation, and then there's the science that we didn't even know existed, until Kim Tingley told us the story of how Micronesian explorers have navigated the Pacific Ocean for generations, without instruments, relying upon their unmatched understanding of ocean waves.

Two thousand sixteen was also a year of record highs: it was the warmest year ever recorded, both on land and within the oceans, its average temperature approaching one entire degree Celsius above the 20th-century average. Not unrelatedly, the carbon dioxide concentration within the atmosphere hit a new high, tipping the 400 parts-per-million mark, possibly for good and all. Only 30.1 percent of Earth's surface was forested as of 2016, probably the lowest value since trees first evolved hundreds of millions of years ago. Global population is at an all-time high, as are crop and livestock production. The proportion of people who live in urban areas has skyrocketed to an unprecedented percentage, and energy use has followed suit. Science journalists spent much of 2016 reporting on how this is changing our planet.

While Elizabeth Kolbert wrote about ice melting in Greenland, Tom Kizzia was documenting the effects of climate change on Alaska's Inupiat peoples. Our urban spaces got some of the attention that they deserve: Omar Mouallem wrote a graceful piece on light pollution, Tom Philpott reported on the factory farms that feed our cities, and Becca Cudmore questioned the role of rats within city ecosystems. Of special interest was Los Angeles, the most densely populated urban area in America: Nathaniel Rich divulged the mystery of its methane leaks, while Adrian Glick Kudler explored the significance of the famous Santa Ana winds. As the twin forces of globalization and industrialism proceed full force, our protected spaces are more important than ever, as Michelle Nijhuis illustrated with her piece on national parks. Picking up the theme, Christopher Solomon shared with us the myriad disagreements over protected land use in the American West, and Robert Draper told us the dramatic story of Virunga National Park, located in the Democratic Republic of the Congo.

Because 2016 was also the year that my own book, *Lab Girl*, was published, I was sensitive to stories that illuminated the life of the individual scientist and showed all the different ways that science can be practiced. Emily Temple-Wood enlightened us about several ancient and overlooked women scientists who should rightly take their place as role models for a new generation of students, while Chris Jones profiled modern-day astrophysicist Sara Seager. Several journalists told us the stories of scientists who did the unexpected, from Sonia Smith's profile of Katharine Hayhoe and her quest to convince Christian evangelicals of the data demonstrating climate change, to Sally Davies on former physicist Fotini Markopoulou and her decision to leave everyday academia, to Michael Regnier's report on the interesting story of George Price and his obsession with altruism. Some journalists went a step further and broke down the very idea of what it means to be a scientist, such as David Epstein on the "do-it-yourself" science of genetics and family history and Jon Mooallem's thoughtful piece on the "amateur" cloud scientists who changed the field of meteorology. With her tongue placed firmly in her cheek, Ann Finkbeiner told us the real story of how "starshot" science gets funded when wealthy investors team up with overconfident experts.

Scientists and readers alike were obliged to continue questioning whether science is a place where women can thrive, after multiple lawsuits, investigations, and resignations associated with sexual harassment surfaced at high-profile institutions, including the University of California at Berkeley, the University of Chicago, the American Museum of Natural History, and the University of Washington at Seattle. In recognition of her fearless reporting in this area, I've made special inclusion of the article by Azeen Ghorayshi describing the incidents at CalTech that led to the unfair firing of a graduate student, as well as Kathryn Joyce's piece exposing the infuriating reality that women are not safe while working within U.S. national parks and forests.

No one can argue that 2016 wasn't a busy year: the scientists of the United States produced more than 400,000 (abstruse) journal articles, and the overworked science journalists of this our Internet Age diligently searched out, or perhaps accidentally stumbled upon, the most arresting, intriguing, moving, beguiling of the bunch and then spun them up into stories for websites, magazines, and newspapers. Here I offer what I believe were the best of the best, the pieces that illuminated science as both glorious and tragic and shined a light on a great discipline that fosters both the best and the worst of what our institutions can be. The year 2016 proved to us once again that science is both essential and frivolous, jubilant and despairing, lovely and brutal, perfect and broken—all at the same time—just like the scientists who fashion it.

HOPE JAHREN

PART I

Emergent Fields

SARAH EVERTS

The Art of Saving Relics

FROM *Scientific American*

THESE SUITS WERE built to last. They were pristine white and
composed of 20-plus layers of cutting-edge materials handcrafted
into a 180-pound frame of armor. They protected the wearers
from temperatures that fluctuated between –300 and 300 degrees
Fahrenheit and from low atmospheric pressure that could boil
away someone's blood. On a July day in 1969, the world watched
intently as astronaut Neil Armstrong, wearing one of these garments, stepped off a ladder and onto a dusty, alien terrain, forever
changing the landscape both of the moon and of human history.
Few symbols of vision and achievement are more powerful than
the Apollo mission spacesuits.

Back on Earth, the iconic garments found new lives as museum pieces, drawing millions to see them at the National Air and
Space Museum in Washington, D.C. And staff members there have
found, to their surprise, that the suits need their own life support.
They are falling apart.

Last year Lisa Young, a conservator at the museum, noticed that
a white, foggy bloom was beginning to creep across the transparent fishbowl helmets and that their smooth, curved surface was
beginning to crack. "It is really frustrating," Young says. "We had
thought they were relatively stable." There had been warning signs
of suit trouble, though. The neoprene pressure bladders that kept
astronauts' bodies from exploding in the vacuum of space began
crumbling years ago, releasing acidic gases. "Anybody who has
worked with the spacesuits knows their smell," Young says. "I'd

describe it as slightly pungent sweet chlorine." And an orange-brown sticky stain began appearing on the exterior white fabric.

The trouble is the construction material: plastic. Most people think plastics last forever, which makes them a bane to the environment. But although the repeating units of carbon, oxygen, hydrogen, and other elements in plastics have a long lifetime, the overall chains—synthetic polymers—do not age well. Light conspires with oxygen and temperature to weaken the bonds that hold the units together. Then chemicals added to plastics to make them bendable or colorful migrate outward, making the surface sticky and wet and perfect for attracting dirt. The polycarbonate spacesuit visor, Young thinks, was leaching out a substance added to make it easier to shape.

Priceless 20th-century art is in serious trouble as well. In that era, Andy Warhol, David Hockney, and Mark Rothko all used acrylic paint—a plastic polymer popularized in the 1940s as an alternative to traditional oil paint. Plastic is, in fact, a building block of much of our recent cultural heritage, including important designer furniture, archival film, crash-test dummies, the world's first Lego pieces, and Bakelite jewelry, as well as the plastic sculptures made by the pop-art movement. "We now know that objects made of plastic are some of the most vulnerable in museum and gallery collections," says Yvonne Shashoua, a conservation scientist at the National Museum of Denmark and one of the first cultural heritage researchers to study plastic degradation.

The conservation field is now racing against time, trying to keep pace with the material's unexpectedly rapid deterioration. Conservators have identified the most trouble-prone plastics. Scientists are developing new tools to diagnose plastic degradation before it becomes visible to the human eye—for example, by measuring the molecules wafting off artifacts. Researchers are also devising new strategies for freshening up precious plastic art without harming it, using everything from cleaning solutions called microemulsions to polyester microfibers that gently remove dirt.

Degrading Denial

The realization that plastics were a problem dawned slowly. For most of the 20th century the museum world was afflicted with "plastics denial syndrome," Shashoua says. "Nobody thought that plastic objects in their collections would degrade." In fact, some conservators were so enamored with plastic during its heyday of the 1950s, 1960s, and 1970s that they used the polymers in ill-advised ways themselves. For example, conservators laminated Belgium's oldest parchment, the Codex Eyckensis from the eighth century A.D., with PVC plastic for protection. Decades later this laminate had to be painstakingly separated from the parchment because changes in the PVC began exacerbating the ancient document's demise.

Crash-test dummies first made Shashoua think plastic was not forever. She had grown up visiting London's Science Museum, where dummies built in the 1970s to better understand the human toll of automobile collisions were on display. The mock bodies—among the first of their kind—have a metal frame skeleton enveloped by medical gelatin that has been sculpted into human form and then covered by a layer of protective PVC. During impact tests, encapsulated red paint would bleed out of the gelatin bodies and get caught underneath the PVC layer wherever the dummy had smashed against a car frame during collision experiments. The red wounds indicated the body's most vulnerable regions.

As the decades passed, these same crash-test dummies in the museum began bleeding again. Shashoua was shocked to see that the PVC covering these artifacts was collapsing, dripping so much wet, sticky muck that museum staff had set up petri dishes in the showcase to collect the mess. When Shashoua was put in charge of cleaning the artifacts in 2011, she noticed that the dummies' sculpted contours were losing their definition as the PVC plastic collapsed; in some parts the red paint mixed with the wounded plastic, giving the goo dripping from the dummies an eerily realistic brownish red tinge.

This dripping mess—and in fact all kinds of plastic degradation —owes its start to oxygen. With help from light and heat, the gas rips off the electrons from the long polymer chains that entwine to form a plastic object. Losing electrons can weaken and break chemical bonds in a plastic, undermining its structure. Essentially the long chains break up into smaller constituent molecules called monomers. In the case of the crash-test dummies, this destabilization allowed ingredients called plasticizers, which are added to make the plastic supple, to pour out.

When the museum world began to realize that plastics were not invincible to time, those tasked with protecting plastic art and artifacts had to start from scratch to understand in detail why their collections were breaking down, says Matija Strlič, a conservation scientist at the Institute for Sustainable Heritage at University College London. Although there was extensive literature on polymer production, this research stopped at the end of a plastic object's expected lifetime—right when conservators get interested, Strlič says. Polymer makers had probably expected that old plastic objects would get tossed away, not delivered to museums.

The Feared Four

Conservators learned that four kinds of plastic polymers are especially prone to problems: PVC, found in everything from spacesuit life-support tubing to crash-test dummies; polyurethane, a primary ingredient in products as diverse as pantyhose and packing sponges, as well as sculptures made from these materials; and finally cellulose nitrate and cellulose acetate, two of the world's first industrially produced synthetic polymers, found in the film used in early cinema and photography as well as in artificial tortoiseshell items, such as vintage combs and cigarette holders.

Cellulose acetate and cellulose nitrate are not only fragile, they are also often referred to as "malignant" by conservators, Shashoua says. That is because they spread destruction to nearby objects. As their polymer networks collapse, they release nitric acid and acetic acid as gases. (Acetic acid is what gives vinegar its characteristic smell and degrading film an odor reminiscent of salad dressing.) The acids eat away at objects made of these plastics. To make matters worse, their gases can also corrode metal and textile things in the same display case or nearby storage. That smell of vinegar is an alarm bell not just that these objects are destroying themselves but that the degrading polymer is taking down innocent bystanders as well.

Shashoua has seen fashion display cases where the acids from a degrading plastic comb have begun eating away textile outfits showcased with the comb or where the plastic in faux tortoiseshell eyeglass frames releases acid that corrodes the spectacles' metal hinges. Once, in her own workspace, a box containing knives with cellulose nitrate handles began releasing nitric acid that corroded both the metal blades and the hinges of a cupboard near where the utensils were being stored, Shashoua says. To stop these chemical attacks, conservators may put objects made of cellulose acetate in well-ventilated spaces to whisk away the dangerous gases. They also capture the poisonous gases in the tiny pores of filters made from activated carbon and zeolite, in much the same way gas masks protect troops exposed to chemical weapons.

Ventilation and trapping are good strategies against cellulose acetate and cellulose nitrate, but the methods do not work on all plastics, Shashoua says. For example, when PVC breaks down, if its degradation products are pulled away from the surrounding environment, the plastic just releases more. Instead conservators need to keep PVC locked down, sealed in airtight containers, to stall its demise. When conservators noticed that the pristine white Apollo mission spacesuits were getting orangey brown stains on their nylon exterior, they realized the cause was plasticizer leaching out of life-support tubing made of PVC that had been sewn into the textile. The tubing kept astronauts' bodies from overheating by circulating cooled water around the outfit. "We had to carefully remove all the life-support tubing from all the Apollo suits and store it separately in sealed containers," Young says. "That was a lot of work."

These opposing approaches—sealed containers versus ventilated ones—highlight why there is no one-size-fits-all solution. "No two objects are alike," Strlič says. For this reason, conservation scientists try to identify the base polymer in a plastic artwork or artifact, typically with analytical machines such as a Fourier transform infrared spectrometer, which bounces long wavelengths of light off an object to reveal its unique molecular fingerprint. Conservators at the Solomon R. Guggenheim Museum in New York City used such a method to uncover a hidden danger in artwork by Bauhaus pioneer László Moholy-Nagy. They had believed the base material for his painting *Tp2* was Bakelite (a phenol-formaldehyde resin), says Carol Stringari, head of conservation at the museum. But recent infrared spectrometer analysis by scientists affiliated with the Art Institute of Chicago revealed that the polymer was actually cellulose nitrate, one of the plastics that can release harmful gaseous acids.

Spectrometry used in this way is helpful, but it has limits. It can identify many ingredients, but it does not always show the entire potpourri of dyes, stabilizers, surfactants, plasticizers, and antioxidants that are mixed into plastics. Often industrial manufacturers keep these recipes secret as part of their intellectual property. Because there is no easy reference for their components, it requires arduous analysis to uncover the plastic's chemical makeup.

These additives change the way an object will age and fall apart. Some varieties of PVC, such as the kind in the spacesuit's life-support system, break down by leaching a sticky plasticizer called di(2-ethylhexyl) phthalate. Other PVC objects degrade by developing a white, powdery crust on the surface: in this case, stearic acid is to blame. It is a lubricant added to the plastic to prevent the polymer from sticking to its mold during the manufacturing process.

Sniffing Out Decay

It is so important to identify the chemical mélange before developing a life-extension strategy that researchers are literally sniffing out the ingredients in plastic artifacts. For example, in a project aptly named "Heritage Smells," Katherine Curran of University College London capitalized on the fact that a lot of degrading plastics emit stinky molecules. Not only does cellulose acetate smell like vinegar as it breaks down, and aging neoprene like sickly sweet chlorine, but many other plastics also release volatile molecules as they disintegrate: degrading PVC has the aroma of a new car, and degrading polyurethane can smell like raspberry jam, cinnamon, or burning rubber. These are just the odors detectable by the human nose. Curran developed a mass spectrometry technique that analyzes all the volatile molecules rising off plastic objects to pinpoint the additives and stabilizers breaking down in a plastic. The goal is to identify what is going on inside without needing to take a sample and to do so before there are visible signs of decay, Curran says.

Curran took her technique to the Birmingham Museum & Art Gallery, where she sampled the air around an enormous art installation made in 2005 by Benin artist Romuald Hazoumé called *ARTicle 14, Débrouille-Toi, Toi-Même!* which translates to *ARTicle 14, Straighten Yourself Out, by Yourself!* It features a market cart full to the brim with sports shoes, computers, a film reel, golf clubs, old

Nokia phones, toys, pots, pans, high-heeled pink shoes, and a vacuum cleaner, to name just a few components in the piece, which Hazoumé put together from objects he had collected during the 1990s and 2000s. Amid the chaotic artwork, Curran and her colleagues detected the presence of acetic acid, one of those corrosive gases that can hurt nearby materials. "We found that the film reel—specifically, degrading polyester in the film—was emitting the acid," Curran says. Museum staff are now considering whether to store the film reel separately or use absorbents for the acid to prevent it from having a detrimental effect on other components of the piece, she says.

Curran has also tried out her canary-in-a-coal-mine technique at the Museum of London on a collection of vintage handbags —purses made of faux leather, mock tortoiseshell, or coiled 20th-century telephone cords. In the case of the white-telephone-cord purse, Curran sniffed out the presence of plasticizers that typically emerge from degrading PVC—a useful alarm bell for staff, who may want to store the purse in a sealed container.

Researchers are also turning to new imaging technologies that create detailed two-dimensional maps of the chemical composition of an object, essentially going pixel by adjoining pixel. For example, Strlič has combined near-infrared spectroscopy with a digital camera to produce two-dimensional colored maps from which conservators can identify the molecular makeup of artifacts that contain many types of plastic, as well as the migration of degradation chemicals. Strlič has gazed inside a popular vintage piece from the 1950s called a crinoline lady—where a plastic bust of a woman forms the handle of a hairbrush. Strlič's team used the technique to identify the handle as cellulose acetate and the brush hair as nylon, using color gradients to show the location of the two plastics in the artifact. By identifying potential dangers such as the acetate, museum staff might be able to take action before damage is visible to the naked eye.

Although researchers are getting better at diagnosing how a plastic artifact or artwork is degrading, they are still trying to figure out how to best stop the decay and repair damage. That was one challenge tackled by a project called POPART, or the Preservation of Plastic ARTefacts in Museum Collections, which started in 2008 and combines efforts from institutions around the world. Cleaning may make the object look better, but it might eventually accelerate the overall demise. A white crust on the surface might be unsightly but is also a protective patina, similar to the green oxidized layer that forms over aged copper as both a degradation product and a protective skin.

Cleaning Up

Even if washing off this patina is the right strategy, POPART researchers want cleaning methods that can do so safely. Conservators are very cautious—a good characteristic in those charged with caring for million-dollar art. And plastics can get cracked, dissolved, or discolored when exposed to the wrong cleaning agent. POPART investigated approaches ranging from high-tech microfibers and ultrasound to carefully formulated cleaning microemulsions (solutions of water, oil, and a surfactant that lifts dirt), as well as gels. The scientists learned that cleaning a polystyrene object with acetone—often used in nail polish remover—could turn the plastic from transparent to opaque and eventually dissolve it. Isopropanol, a different alcohol-based cleaning solvent, however, is safe for most plastics.

Using something as simple as water to clean acrylic paintings turns out to be risky, says Bronwyn Ormsby, a conservation scientist at the Tate, a group of four museums in England. She confronted that problem with the 1962 painting *Andromeda,* the Tate's oldest acrylic piece. Russian-American artist Alexander Liberman painted this abstract, geometric work on a circular canvas; its four solid colors—black, lilac, dark purple, and dark green—evoke the darkness of outer space. But acrylic paints have additives called surfactants that help to keep pigments suspended in the paint tube rather than settling to the bottom. That is good for the painter. Yet once on a dried canvas, these surfactants migrate to the surface and create a sticky substance that attracts dirt. By 2007 *Andromeda* was obscured by so much surfactant buildup that the painting had "a whitish bloom, which is quite distracting on paintings with dark colors," Ormsby says. Ordinarily she would turn to water as a cleaner: "Water often removes soil better than any other solvent." But water also makes acrylic paintings swell. That can lead to a loss of paint during the cleaning process.

Water can be tweaked to make it safer, though. Investigators led by Richard Wolbers of the University of Delaware have found that keeping water's pH levels around 6 and making the water moderately salty can limit the swelling of acrylic paint. Ormsby used that technique on the Liberman painting, which today looks as dark and lonely as it did five decades ago. Researchers at the Tate have also used an atomic force microscope to monitor Warhol's acrylic portrait of Brooke Hayward as it was cleaned, to make sure dirt and not paint was being removed.

Sustainable Art

Ormsby and others are also working with scientists at Dow Chemical to use the company's industrial-scale abilities to run a large number of chemical reactions quickly to test a variety of micro-emulsions on acrylic paint samples. Their goal is to try different combinations of cleaning compounds to find the best formula for washing painting surfaces without harming them.

Plastics researchers are also reaching out to artists to let them know about the potential pitfalls of producing art from plastic. "The idea is not to interfere with the creative process but to allow the artists the option to use this information if they wish to," says Carolien Coon, who is an artist herself, as well as a conservation scientist at the UCL Institute for Sustainable Heritage. Coon says she wonders about a sculpture she sold years ago that was made of silicone rubber, a bronze cast, a fishbowl, and baby oil. "I have no idea how it looks today. I hope it hasn't leaked all over the dining room table."

The great hope of conservation scientists is that restoring the past will also help them prepare for the future, when today's plastic materials—such as 3D-printed objects—start entering museum collections. One such item might be the first 3D-printed acoustic guitar or a retired International Space Station suit. Eventually all will be past their prime, and conservators want to have the tools in hand to give these cultural icons a facelift.

MARIA KONNIKOVA

Altered Tastes

FROM *The New Republic*

THE LIGHT IN the room softly brightened and grew warmer, yellower, somehow more embracing. A quiet rustling—wind through leaves?—reached my ears. A white mist covered the table, carrying with it, somehow, the smell of damp earth after a late summer storm, and the promise of the mushrooms which would bloom in its wake. At the center of my table: a cylindrical terrarium-like enclosure filled with layers of soft green moss, soil, and broken branches, complete with a miniature tree. A plate was silently placed in front of me, or rather, a dark brown platform of what looked at first to be sod (actually a mixture of beetroot and mushroom powder with truffle), adorned with bursts of yellow pollen (a compact butter with truffle, root vegetables, and salt), anchored by a crinkled log (potato-starch paper covered in smoked salt, powdered mushroom, and porcini), punctuated by tiny green leaves (fig leaves), and at the bottom a thin layer of mushrooms (button, anchored by a mushroom stock jelly). Beneath all this theatricality was an undeniably delicious dish. Even today I recall its flavor and think of these as the best mushrooms I've ever eaten, though in fact I've consumed ones both more rare and more expensive. A map, which serves as a sort of menu for the Fat Duck, Heston Blumenthal's three-starred Michelin restaurant in Bray, England—one of only four in the country—described this dish as "damping through the boroughgroves." Presumably no mome raths would be consumed.

It was mid-October, and the Fat Duck, one of England's best-known modernist restaurants, had just reopened after a nine-

month hiatus and a \$3.6 million redesign and reconceptualiza-
tion. "It was time for a change," the 49-year-old Blumenthal told
me as we walked the restaurant. Blumenthal—one of the leading
chefs in the world, a constant presence on global best-of culinary
lists, and the host of several television shows in England—is a towering figure in his chef's whites and thick, black-framed glasses,
equal parts cook, linebacker, and fast-talking salesman. It's difficult not to get swept up in his exuberance. *Of course* it was time
for a change, and *of course* the changes were *incredible.* "We'd been
going so fast for so long that we couldn't keep up with all of these
exciting advances happening outside the kitchen. It was time to rethink." And so, while a pop-up carried on the restaurant's name in
Melbourne, Australia, Blumenthal and his team continued working, only instead of serving customers they invented new dishes,
conducted experiments, and devoted themselves to creating a
dining experience on the frontier of gastronomic science, a place
where brain, body, and nutrition intersect.

The Fat Duck's map wasn't a menu in the traditional sense. It offered few clues to each dish's contents, and the dishes themselves
often appeared to be more Wonderland-like flights of whimsy than
actual food. I was treated to Mock Turtle's soup (complete with
the White Rabbit's gold watch); March Hare's tea (the "itinerary"
included with Blumenthal's map reads, for this dish, "Excuse me,
there seems to be a rabbit in my tea"); and a (literally) floating
dessert concoction marked "counting sheep." Lewis Carroll served
as an inspiration, Blumenthal said, a means to stimulate the "fun"
and "curiosity" that he feels should be part of any food experience.
But the intention behind the food was quite serious. Blumenthal
was trying to persuade people to engage with food in a fundamen-
tally new way, one that is both physiological and emotional. "It's
not just about the food," he said. "It's the ebb and flow of the story,
the look and feel of the room, the temperature, all that."

The Fat Duck is operated along the principles of neurogastronomy, an emerging scientific field that examines how our sense of
taste is interpreted and reinterpreted by the brain. The term itself
was coined about a decade ago by Gordon Shepherd, a neurobi-
ologist at Yale, who has been studying the science of olfaction for
more than half a century. His research has shown that flavor, a
complicated and little-understood concept, does not originate in
what we eat but in what our minds derive from the experience.

"Our sensory and motor appreciation of what we have in our mouth is created by the brain," he said. "We can't have gastronomy without it."

This is the overarching principle that guides neurogastronomy: what we eat and why we eat it is as much a psychological phenomenon as a physical one. Throughout most of history, eating has been understood as a primitive human characteristic, an evolutionary necessity, the stuff of base survival instinct. This perception turns out to be far too simplistic. The more we learn about flavor, the more we realize just how easy it is to manipulate. Not just by the overclocked sensations of processed food, but in ways that make healthier choices seem at once tastier and more satisfying. Though most of us would like to think we have discerning palates, our taste is quite easy to fool.

When we try to imagine the flavor of something, we tend to focus on our mouth—the experience of placing, say, a ripe strawberry on our tongue. But that in fact is *taste,* and though we tend to conflate it with *flavor,* a vast chasm exists between the two. Taste is an experience composed of only five elements: sweet, salty, bitter, sour, and umami. Thousands of receptors on our tongue are designed to identify and respond to these elements, each one specializing in one of the five qualities. Without input from other senses—most notably our nose, but also our eyes, ears, and even hands—taste is merely a flat, single-note sensation with none of the nuance or enjoyment we associate with food in general and with specific foods in particular. Flavor is at once a broader and more powerful property than taste, one that marries the senses and their associate properties—memory, experience, neurobiology—to create and control the way we eat.

The promise that neurogastronomy holds is that once we understand how the mind combines the disparate biological and evocative forces that create flavor, we will be able to circumvent the learned and innate preferences of our taste buds. And with that capacity—truly an example of mind over matter—instead of stimulating appetite via the conventional and unhealthy trifecta of salt, sugar, and fat, we can employ the neural pathways through which flavor is constructed in the brain to divert attention to different, more nutritious foods. Control flavor and you control what we eat—and perhaps, given time and more research, begin fighting the global nutrition problems that are a direct result of the industrialized production of food.

Our preferences for salt, sugar, and fat evolved within the context of our species' historical nutritional scarcity. These basic tastes are the echoes of prehistoric signals that saw humanity through epochs of less abundant food sources. They made sense when we were hunter-gatherers eating only what we could kill, less so when navigating the line at the local Subway. Indeed, our basic physiological response to taste is largely innate. Give an infant something sweet and she will lick it up. If it's bitter, she will spit it out. (Bitterness signals potential poison.) We learn to like certain complex tastes over time, but our cravings for sweetness and fattiness remain constant. And so we continue to consume and store reserves for a hard winter that, today, never comes.

Ivan de Araujo, a neuroscientist at Yale Medical School who studies energy and reward in the brain, calls this the great conundrum of humans and food. "Why do we tend to violate homeostasis and equilibrium and eat more food than we need physiologically?" he asked me recently. "Why is there a bias in getting more energy than you're going to expend?" The genetics of weight gain, psychological traits, and sensory perception of food are informed by these questions. When we attempt to address problems of global nutrition, we fight an uphill battle against the energy-craving and -storing machine that is the human body.

These signals, rooted in evolutionary biology, have given rise to a paradox of malnourishment amid a global abundance of food. In many countries, including England and the United States, poor diet now rivals smoking as the greatest public health risk. Malnutrition does not necessarily mean lack of food, but rather lack of proper nutrients. You can eat five meals a day and qualify as malnourished. (Case in point: Morgan Spurlock's near-lethal experiment in *Supersize Me*.) When it comes to certain nutrients, in fact, an estimated 80 to 90 percent of obese individuals are malnourished. (The same percentage holds for nonobese individuals.) Globally, more than 600 million adults and 42 million children under the age of five are obese. Alongside the rise in weight, we've seen a corresponding increase in diabetes, heart disease, and a host of other diet-related problems. In the United States, more than a quarter of the population suffers from some form of metabolic syndrome or illness. Nutrition in many ways is the great public health battle of our times.

It's more than just the descent of man, of course. Modern society often perceives health and flavor as mutually opposed. In England, I met Jozef Youssef, an energetic 34-year-old chef who splits time between research collaborations with psychologists and running Kitchen Theory, a London pop-up restaurant where he tests various scientific findings on small groups of diners—including ways to make "healthy" and "tasty" seem complementary rather than antagonistic. Youssef cited the example of a green smoothie. "We see it and we think, it's good for you but it's probably not enjoyable," he said. "Why is that?" (This also works the opposite way lately, as people have been questioning exactly how healthy the overpriced green smoothies are at Whole Foods.) A study published this year in the *Journal of the Association for Consumer Research* found that people tend to feel less full and eat more after consuming a food they perceive as "healthy," even if it's identical to one that is marked as unhealthy. For example, they will feel hungrier after a "healthy" cookie—and go on to eat more overall.

In this way evolution and socialization are locked in unending conflict, nature and nurture conspiring to produce ever-increasing consumption of less-than-ideal foods. And indeed, past efforts to replace the salt-sugar-fat trifecta with healthier equivalents haven't been successful. Think margarine instead of butter, saccharin replacing sugar, artificial polymers (macromolecules synthetically created in a lab) replacing the fat content of milk or ice cream. It's your low-fat frozen yogurt, diet soda, the 100-calorie snack pack. The results, from a flavor perspective as well as a dietary one, have been underwhelming. Instead of curbing obesity and metabolic disorders, these innovations seem to have resulted in the opposite. This may, of course, be correlation rather than causation, but still —and perhaps worse—some of the substitute substances haven't proven to be as healthful as first suggested. Companies scrambled to scrub margarine from their recipes after research showed that it was actually more harmful than the butter it replaced, while recent work from people like Dana Small, a neuropsychologist and physiologist at Yale, highlights the metabolically disruptive effects of artificial sweeteners, even ones derived from natural substances, like Splenda.

What's more, many (most? all?) low-fat and low-sugar products don't taste good, aren't eaten as often as their sinful counterparts, and end up a bust both nutritionally to the customer and financially to the producer. And the consumption of the real thing keeps rising. Sugar was first introduced to the Western palate via New Guinea about 10,000 years ago. By 1800, Americans were consuming an average of seven pounds of the powder a year. Today our consumption tops over 100 pounds. (By way of comparison, we eat about 50 pounds of beef.) Though we are drinking less soda than before—2012 production was 23 percent lower than a decade prior—people are still taking in 30 gallons of regular soda per person each year, according to New York University professor and public health advocate Marion Nestle. And things like diet soda seem to have a reverse psychological effect: new research suggests that tricking your brain into thinking it has consumed calories when it hasn't can, over time, have a host of negative metabolic consequences, as the connection between the energy signal of sweetness and its actual energy content decouples. And so you consume more, and even worse, you *want* to keep consuming more (the dreaded sweet tooth).

Why should neurogastronomy be different? Why would it succeed where basic physiological nutrition has failed? Part of the answer stems from the insight research has given into how exactly our bodies derive energy and flavor signals from food via the brain. It's not about calories—that is, eliminating calories in the manner of artificial sweeteners likely won't work. Instead it's a far more complex process of taste perception. It's a growing understanding that psychology plays a more central role in the experience of eating than previously thought, a realization that we need to be fooling the brain, not the body.

In 1936, H. C. Moir, an analytic chemist from Scotland who had worked at a baked goods factory, presented what may be the earliest findings that show just how much our brain affects taste. He had people eat incongruously colored jellies—green-colored orange, red-colored lemon, and the like. He then had them taste chocolate-colored sponge cakes, one of which was imbued with cocoa and the other with vanilla. Only one person was able to correctly identify each taste in the two tests—and over half got more than 50 percent of the answers wrong. Some of the answers given for the orange candy: almond, strawberry, blackcurrant, and pineapple. For lemon: cherry, raspberry, strawberry, damson. Some tasters thought the chocolate biscuit was vanilla, and the vanilla chocolate, while others volunteered coffee, orange, or even unflavored.

"The majority of those who came below 50 percent went to great pains to assure me that they were considered by their wives or mothers, or other intimates, to be unduly fastidious about their food, and were invariably able to spot milk turning well in advance of any other member of the household," Moir wrote. "Consequently, it was obvious that the method of testing was at fault, and not the palate being tested. Further, many brought in a plea of individual idiosyncrasy in that they did not like table jellies etc., but comparatively few made this plea before the test."

When it comes to the caloric content of food, our brains aren't easily fooled. You can, it turns out, engineer all the low-fat polymers and artificial sweeteners you want, but they will likely not make us eat fewer calories or gain less weight: our brain is too smart for that. In one study, de Araujo genetically engineered a group of mice so that they would no longer taste sweetness. The receptors that signal "sweet" to the brain simply didn't function. He found that though they started off indifferent to sugar, the mice soon learned that when they were hungry, it was better to consume a solution with sugar rather than one that was all water. They had no way to distinguish the two from a sensory perspective—tastewise, to them, they were identical—but somehow their brains learned where the energy source lay. Soon the rodents were consuming just as much sugar as nonmodified animals. The effect was completely absent with artificial sweeteners. In another study, de Araujo followed up with a group of regular mice. This time the mice were offered two sweet solutions, one with sugar and one with artificial sweetener. The solution with the sweetener tasted sweeter—and so one would think would be the more attractive of the two. And indeed, for the first day the mice consistently drank the sweeter water. But then something happened: they began to ignore the artificial sweetener and instead focused exclusively on the real sugar solution. "Somehow the brain knows when something is purely sweet and good-tasting versus when that good taste comes along with energy," de Araujo said. In other words, we have two separate systems that signal the value of food. It's not just about taste; it's about how taste is incorporated into our brain's reward system. And artificial substitutes—ways of lowering calories while keeping their sensory qualities—simply do not work. The brain isn't fooled. It knows real calories from the taste of calories.

Research like de Araujo's doesn't just show us what won't work: it suggests what might work instead. The deeper understanding of sensory integrations gives an alternative approach to making nutritional changes to the diet: alter rather than substitute. Use real sugar, real energy, real fats and salts and the whole gamut of flavor, but do so in lower quantities, in a way that makes the result taste good and sends actual energy signals to the brain, creating an experience that is both psychologically and physically satisfying.

On my wander through the boroughgroves at the Fat Duck, my dish wasn't just a plate of food. It came with an abundance of theatrical effects, all of which served specific functions. My hearing was engaged—the crackle of the ground and rustle of leaves. My vision—not just the beauty of the plate but the forest mist, the mini-terrarium, the variegated effects of the lighting. ("Unique to each table," Blumenthal pointed out. Each diner's lighting is calibrated depending on where she finds herself on her journey at any given point.) My smell—both when I first inhale the earthiness (via orthonasal smell, or what we typically think of as smell) and after I put the first bite of the powdered-mushroom log into my mouth and exhale (by way of retronasal smell, or, as Shepherd explained, "the smell that comes internally, from behind, from our mouths into our nasal cavities"). My touch—the texture of the smooth mushroom, contrasting with the roughness of the faux bark, the crispness of the greens, the creaminess of the truffle butter. Even something I don't usually think of as a sense—my memory—combined to put me in mind of memories of family walks through the woods. The idea is that due to the multisensory pleasure of the experience, not only will I enjoy the dish more, but I will feel more satiated after having eaten less. "When you first see it you think this guy's taking the piss," said Francis McGlone, head of the Somatosensory and Affective Neuroscience group at Liverpool John Moores University and until recently head of the food neuroscience research group at Unilever. "There's nothing on the plate but these small portions. But you won't leave the restaurant hungry. Because there's so much complexity, in terms of textures, colors, tastes; it's almost symphonic. You reach satiety faster."

Fast food is so addictive because salt, sugar, and fat never appear together in nature. Try to imagine a naturally occurring food that is fatty, has high amounts of sugar, and is salty to boot and you'll come up short. And so, strongly reinforcing neural pathways that were only ever meant to fire in isolation, to tell us that a food is worth eating, now activate all at once, creating an enticing, addictive cascade that is greater than the sum of its parts. In a sense Blumenthal's approach is doing the same thing—only using complex psychological flavor rather than purely physiological taste reinforcement.

One basic approach to this focuses on actual changes to the composition of food—ways of cooking different substances that make them taste sweeter, saltier, or spicier than they actually are. A favorite of Blumenthal's, which he has used successfully in the Fat Duck for more than a decade, is a method called encapsulation, in which he presents a flavor in a way that makes it seem far larger than it is. "If you think of a cup of coffee made with one ground bean, that would be really insipid," Blumenthal suggested. But if you crunch a bean in your mouth, you suddenly have a much stronger coffee flavor, even if the cup itself is quite weak. A single whole bean can deliver a greater flavor punch than multiple beans that have been ground and brewed. The same thing happens with a spice, like coriander. Add a few seeds to a dish rather than grinding them, and suddenly the flavor becomes much more intense, even though the overall quantity of coriander goes down. "Every so often your mouth crunches one and there is an explosion of flavor that makes it much more interesting," Blumenthal said.

With an encapsulation approach—a few strong bursts rather than dispersed flavor—Blumenthal has successfully reduced the salt content of multiple dishes in his restaurants. The final taste experience is just as salty overall, even though the amount of sodium has been reduced. It's a method that one could see playing out in mass-produced items, including your beloved Doritos, fast-food fries, snack bars, cereals, even packaged meals that rely on large doses of sodium to deliver post-frozen taste. "You have fewer, larger grains of salt, and suddenly you can deliver the same flavor, but with less," Blumenthal told me.

Barry Green, formerly at the Monell Chemical Senses Center in Philadelphia and now at Yale, accomplishes something similar through heat. A psychophysicist—someone who studies how physical sensations get interpreted by the brain—Green has worked on the way the thermal sensitivity of the mouth—that is, our perception of hot and cold—can affect a food's flavor. Temperature, and specifically temperature change, can influence how sweet we think something is. Fifteen years ago, while Green was working at Monell, a University of Michigan study found that some of the taste fibers (specialized nerve cells) in the chorda tympani nerve, one of the three cranial nerves responsible for sending taste sensations to the brain, were sensitive to temperature. The nerves responded to warm liquids much as they did to sweetness—even when there was no sugar present.

In a later study Green called this phenomenon "thermal taste": temperature that evokes a flavor. We have fibers that get excited when we warm them up, which might make a food or liquid taste sweeter than its sugar content warrants. One way to think about it is to picture yourself licking an ice cream cone. The initial cold taste isn't nearly as sweet as the flavor of the warmed ice cream once it's back in your mouth. That finding has immediate implications for a multitude of foods. One easy way to get a sweet kick: take sodas, juices, fruit, and whatever else out of the refrigerator.

One of Blumenthal's signature dishes at the Fat Duck is a rabbit "tea"—actually a velouté of rabbit—that is both hot and cold. A specially engineered gel keeps the two sides separate until poured. The result is disorienting: your tongue is hot and cold at the same time. The flavor is intense, the pleasure surprising—and the relatively lower levels of seasoning needed to deliver a flavorful experience perhaps most surprising of all.

While it's difficult to imagine the packaged-food equivalent of dual-temperature tea, the same effect, Green pointed out, can be attained by warming something that is cold or by cooling the tongue itself so that the same food tastes relatively warmer. Two things happen physiologically when the sensation of either warming or cooling hits the tongue. In the first case sweetness increases, while in the second the perception of saltiness becomes more intense. In 2010, Campbell's Soup had something of a PR disaster when it announced that it had lowered the sodium content of its soups. Customers protested that the reformulated versions didn't taste as good, and sales fell. Yet imagine the exact same reformulation but with the introduction of, say, a chilled soup like gazpacho. The cooling would create an enhanced salty flavor, and the soup might replace less healthful alternatives. Knowing some of the thermal principles involved in flavor perception may enable Campbell's to create tweaks not just there but to its hot soups, in ways that reduce sugar and salt but enhance flavor.

Another, potentially broader area of experimentation comes from olfaction. Our brains form associations between smells and tastes that in turn affect both how much we like a certain food and our bodies' anticipated response to it (how our brain prepares the rest of the system for the calories it thinks it's going to consume). Those associations can then be used to trigger the reward system even when the perceived reward is smaller than the actual one. Take vanilla. Vanilla isn't actually sweet. It's quite bitter. But in the Western world we have come to associate it with sweet foods, and so to us it signals sweetness. When we smell it, our sweet receptors go on high alert—and the food we eat tastes sweeter than it otherwise would.

I have to imagine that some of the pungency and sheer fungal intensity of my mushroom dish comes from the olfactory tricks that punctuated it. The fog that spread over the table wasn't just visual: it spread the scent of the moss. Throughout the meal, I could tell without looking when another table had gotten to this particular point in the dinner. The scent heralded its arrival better than anything else could. The mushroom powder on the plate further reinforced and carried the scent, so that by the time I took a bite, all my taste buds were primed for the resulting flavor. A few weeks later Blumenthal told me he still wasn't completely happy with the dish. "We're still working on creating the perfect smell of the woods," he said.

Part of my pleasure from the mushroom dish doubtless derived from the childhood associations I carry with its taste—mushroom-picking with my grandfather, cooking up big skillets of freshly gathered mushrooms and onions in the early fall with the whole family. (We're Russians, after all.) But to someone for whom that affinity is absent, or even reversed, the techniques could have detracted from rather than enhanced the experience, by concentrating the flavor so intensely. Likewise, to a non-Western palate, even something that seems as straightforward as Blumenthal's proposed vanilla addition might backfire. Some Japanese pickled foods contain an almondlike aroma, for example, while sweet almond desserts are mostly absent. The implication here is that taste-smell associations—and the resulting preferences in food—can be changed from experience. Yes, some taste is innate, but the way we perceive it psychologically is a learned process, one that starts in the womb. In one study newborns whose mothers had eaten food with anise during pregnancy enjoyed its scent more than those with mothers who had not. Children of mothers who drank milk flavored with carrots while pregnant were more likely to eat carrots. Adults, unlike children, are far better positioned to make mindful food choices. The fact that associations between basic tastes and nonbasic smells develop so early could become a powerful way to subtly change preferences along more nutritious lines.

In the 1970s, UCLA psychologist Eric Holman discovered that certain sweetened substances could make rodents prefer certain foods by virtue of their presence. For instance, by adding a saccharin to either a banana- or an almond-flavored solution, he was able to make rats prefer the taste of bananas or almonds, respectively, a process known as "flavor nutrient conditioning." In recent years that work has been picked up with humans. Maltodextrin, a glucose polymer, is imperceptible to most of us. It doesn't taste sweet. In fact, it doesn't taste like anything. For it to activate the sweet receptors in the brain, the body must first break it down into glucose. If we mix it into another food, we don't realize there's a sugar present, but we still develop a preference for that flavor. In one study, people who tasted foods with maltodextrin mixed in would reliably choose the flavor that had been associated with the polymer in subsequent tests. They had been trained to prefer one food over another by a sort of sensory trickery. Imagine dusting a child's broccoli florets with maltodextrin and transforming a disliked vegetable into a favorite. Ethically questionable, yes. But also potentially quite effective at nudging children toward healthier choices at a sensitive period in life when many such choices are first formed. The end result would be a society that makes better, more nutritious choices without seeing them as a necessary evil or sensory tradeoff. Broccoli would be a preferred taste, a food you choose because you've learned to enjoy it and find it inherently rewarding. When you went to reach for a snack, a broccoli crisp would be just as, if not more, enticing as a potato chip. "If we can find out how to do that on a large scale," de Araujo told me, "we could completely change diet."

The sneaky additive approach, though, is not one favored by

chefs like Blumenthal. "We have enough naturally occurring flavors that we don't do enough to exploit," he told me. "Like MSG.
It's an old wives' tale that it's bad for you. It occurs naturally all the
time. Tomatoes. Parmesan. Shiitake. Seaweed." We can use natural
properties to create flavor profiles that are apparent—and make
other foods enjoyable by sheer association. It's an approach that
stems from stimulating other tastes that may then make a food
more pleasurable by proxy. In addition to studying heat, Barry
Green has worked with flavors that could have a similar effect,
namely, menthol and capsaicin. The former gives the sensation
of cooling your tongue. The latter, found in peppers, warms it up.
In so doing it stimulates the pain system—but in a way that can be
pleasurable. Could the addition of foods that work on different
neural channels from sweetness and saltiness but that stimulate
the somatosensory system just as strongly help reduce the need for
things like sugar and salt? "Obviously, chili pepper has become a
huge part of the American diet," Green said. "I'd love to see how
that channel of input could be utilized to increase flavor when
you're decreasing things like salt and sweetness . . . using spice so
you don't have to have a chip that's as salty, for instance."

In Blumenthal's kitchen, such approaches are evident in dishes
like the Mock Turtle's soup from the Mad Hatter's tea party. A
faux-gold watch made from gold leaf covering the equivalent of a
beef and oxtail bouillon cube is placed into a teapot of water, only
to dissolve into beef and oxtail stock, which is then poured over
an "egg" that is actually made of rutabaga and turnip, and also
flavored with mustard seeds and pickled cucumbers. The combination of spice, from both the mustard and the pickling, make the
oft-ignored tubers—not many people crave rutabaga—shine in a
new light, with a complexity of flavor not associated with the uniform blandness of root vegetables. "We really undervalue spice,"
Blumenthal said. "But used the right way, it's eye-opening."

The wine in the Fat Duck's second-floor wine cave is hidden behind a faux bookshelf. The bottles are accessible only if you know
the proper title to tip off the shelves—part of the ethos of playful
curiosity and discovery that permeates the whole restaurant. On
the day I visited, the room was empty but for a small round table in
the center. It was covered with a white tablecloth and several iden-
tical-looking glasses of wine, some white, some red. Blumenthal

was running several experiments based in part on the research of Oxford psychologist Charles Spence, the first of which involved the links between dexterity and taste. Isa Bal, the Fat Duck's sommelier, a dapper-looking man in a dark suit, instructed me to pick up a glass of white wine and take a sip.

"What does it taste like?" he asked.

"Smooth," I replied. "A bit buttery?"

"Now pick it up with your left hand." (I'm right-handed.)

I drank again, and it was like a different wine: sharper, crisper, more acidic. One explanation for this lies in the neural wiring of the dominant versus secondary hand. Our dominant hand is more fluent, which means the signals from it are processed more easily. If the results were strong enough, one might expect future dinners at the Fat Duck to include nontraditionally placed glass and silverware—and wait staff who instruct guests on the proper hand with which to try a certain dish or drink. It's not a stretch to imagine such instructions appearing on food packaging: *Tear open with your right hand and dig in with your left for maximal pop—or make sure to hold with your right hand for the fullest buttery feel.*

Next Isa played a series of musical tracks and had me taste wine against different songs. And indeed, with each track the taste changed. Alongside one, the white was greener, more effervescent. Along another, smoother. The music played in the Fat Duck's dining room has been carefully chosen to match the sensory characteristics each dish is meant to convey. The Sound of the Sea—an ethereal plate of mackerel, octopus, and kingfish covered in a rich seaweed foam and arrayed on a sand-strewn beach (the sand is made from tapioca and miso oil)—was presented with headphones that snaked their way out of a conch shell. An iPod was hidden inside, playing a collection of surf, waves, seagulls, and beach sounds.

Did the music add anything, or was it more theater in an already theatrical meal? In a series of studies to test that, Blumenthal and several collaborators had PhD students eat a similar dish, listening to either barnyard or seashore sounds. They rated their enjoyment levels up to 90 percent higher with the sea soundtrack. I got another taste, so to speak, of the same phenomenon when I took part in some of the ongoing studies at Spence's cross-modal laboratory at Oxford. A candy bar tasted sweeter alongside a piano, a jelly more sour alongside brass. "Music changes the sensory and hedonic experience," Janice Wang, Spence's graduate student in charge of the experiment, explained. "The olfactory nerve is partly connected to the auditory—and the more we learn about how the senses are wired, the more we can change the experience by changing the auditory environment."

A little more than a year ago, the Cadbury chocolate company changed the shape of one of its bars from rectangular to round. "People complained about it being sweeter," Spence said. This was in one sense a marketing error—but it was also a missed opportunity. Could Cadbury's not have reduced sugar at the same time, thereby rendering the reduction imperceptible and thus creating a healthier candy bar? "Reducing the ingredient by some amount and changing packaging to make it neutral in the consumer's perception is a very real goal," Spence said. Other studies show that heavier cutlery or packaging makes a food taste better, that certain colors and contrasts can make it taste sweeter, saltier, smoother, more bitter, more sour, even that the language we use to describe it can make a difference in how it's perceived. One can easily imagine that part of the appeal of Mast Brothers chocolate was in the packaging: what they lacked in flavor they more than made up for in artisanal-seeming wrapping. Perhaps this was also one of the reasons the blowback against them was so harsh. People felt deceived, as, in a sense, they were.

No matter how much we learn about neurogastronomy, though, a disconnect between what's possible and what can actually be accomplished will doubtless remain. Not everyone will be happy that the sugar content of their chocolate bar has been lowered. "Innovation is really slow in food companies. It's really difficult to get things done," Spence told me.

About ten years ago Francis McGlone attempted to bring Blumenthal into Unilever, a leader in the world of FMCG, or fast-moving consumer goods, as a consultant. He felt the company would benefit immensely from the chef's creativity. The collaboration went nowhere. "No one could agree who controlled the intellectual property," McGlone said. "They wanted to nail him down to contain his creativity in ways he couldn't accept. It was a missed opportunity." He paused. "It's very difficult to change the way these large companies go about what they do." Theoretically they may be intrigued by Blumenthal's innovations, by his use of the latest science to craft ways to eat more healthfully. But practically they are beholden to their shareholders. The tolerance for risk and acceptance of failure that marks the Fat Duck's experimentations —new dishes are the result of many spectacular mishaps in the test kitchen—are unacceptable from their perspective.

And yet the mandate from consumers seems to be changing in a way that may force large corporations to rethink their approach. Companies respond to consumer pressure. What matters is what people want as expressed by what they will buy. Food manufacturers aren't our parents, nor are they our doctors. They care about profit, not health. But their profit is dependent on shifting preferences and demands. The mass production of the 1960s and 1970s was about the demand for convenience, freeing the housewife from the slavery of cooking. And from that demand came the packaged foods and snacks that fill grocery stores today. But increasingly consumers are demanding more than simple convenience. They want health too. Doug Rauch, the former president of Trader Joe's, just opened a new store, Daily Table, that he hopes will grow from a single location in Boston to a national chain of supermarkets. It aims to sell nutritious food at the price point of fast food. Bananas for 29 cents per pound. A dozen eggs for 99 cents. PepsiCo has just announced a new vending-machine initiative, to be rolled out to several thousand locations in 2016, which will offer healthier snack options than the traditional soda and chips —Naked Juice, Quaker bars, Sabra hummus cups, and the like. And interest in the frontiers of nutrition science has risen apace. "People don't do anything until somebody else does it, then they all want to do it," Spence said with a laugh. "I had three companies on the phone today. It's been an explosive growth in interest. And my colleagues working in this space would say the same."

One initiative, with the University of Barcelona and the Alícia Foundation, a research center helmed by Ferran Adrià, the chef of the now-closed El Bulli restaurant, focuses on improving nutrition and recovery for children with cancer. Most common treatments, including chemotherapy, create an eating experience in which food takes on a metallic, ungainly taste. Unlike adults, who can override such unpleasant inputs with the knowledge that they have to eat to get better, children will often refuse food altogether rather than consume something they don't like. This team of chefs and scientists hope to use the new understanding of the brain's sensory integration of taste to create foods that would override that aversion, by changing either the perception of the taste or its seeming desirability. "The evidence that we can improve consumption is quite good," Spence, who consulted on the project, told me. "And a lot of hospitals and end-of-life care are now engaged in sensory design related to food. It's an important investment."

Blumenthal told me of a recent project he collaborated on with the National Health Service at the Royal Berkshire Hospital to improve nutrition among the elderly. As we age, our sense of smell dampens—loss of smell sensitivity is one of the earliest signs of dementia—and our desire for healthier foods along with it. There's a reason your grandmother douses everything in salt and prefers simple, strong fat-sugar-salt combinations. But something else rises with age: the risk of cardiovascular disease. And so ways of enhancing flavor that don't necessitate huge quantities of condiments and fats could go a long way toward easing the aging transition. "We wanted to get the elderly excited about food again," Blumenthal said. "And I'm hopeful that it can be done well."

Chefs like Blumenthal are an important first step in the advancement of neurogastronomy. But to realize the full health and nutritional potential of this science, we will need to go much further than test kitchens and high-end dining rooms. Neurogastronomy must be incorporated into mainstream consumption. "It's like Formula 1 racing," McGlone said. "The car is the pinnacle of advanced technology. The efficiency and safety of these engines are at the most advanced. But in three, four years it appears in the average car." The same could be said of neurogastronomical innovation. "It will ultimately become part of the standard. It will be an advanced use of the technology that will ultimately trickle down into the standard food product." The fact that why we eat what we eat originates in the mind rather than the palate is a powerful one. Properly harnessed, it could prove to be the key to succeeding where so many other nutritional interventions have failed.

KIM TINGLEY

The Secrets of the Wave Pilots

FROM *The New York Times Magazine*

AT 0400, THREE miles above the Pacific seafloor, the search-
light of a power boat swept through a warm June night last year,
looking for a second boat, a sailing canoe. The captain of the canoe, Alson Kelen, potentially the world's last-ever apprentice in
the ancient art of wave piloting, was trying to reach Aur, an atoll in
the Marshall Islands, without the aid of a GPS device or any other
way-finding instrument. If successful, he would prove that one of
the most sophisticated navigational techniques ever developed still
existed and, he hoped, inspire efforts to save it from extinction.
Monitoring his progress from the power boat were an unlikely trio
of Western scientists—an anthropologist, a physicist, and an oceanographer—who were hoping his journey might help them explain
how wave pilots, in defiance of the dizzying complexities of fluid
dynamics, detect direction and proximity to land. More broadly,
they wondered if watching him sail, in the context of growing concerns about the neurological effects of navigation by smartphone,
would yield hints about how our orienteering skills influence our
sense of place, our sense of home, even our sense of self.

When the boats set out in the afternoon from Majuro, the capital of the Marshall Islands, Kelen's plan was to sail through the
night and approach Aur at daybreak, to avoid crashing into its
reef in the dark. But around sundown the wind picked up and
the waves grew higher and rounder, sorely testing both the scientists' powers of observation and the structural integrity of the
canoe. Through the salt-streaked windshield of the power boat,
the anthropologist, Joseph Genz, took mental field notes—the

spotlighted whitecaps, the position of Polaris, his grip on the cabin handrail—while he waited for Kelen to radio in his location, or rather what he thought his location was.

The Marshalls provide a crucible for navigation: 70 square miles of land, total, comprising five islands and 29 atolls, rings of coral islets that grew up around the rims of underwater volcanoes millions of years ago and now encircle gentle lagoons. These green dots and doughnuts make up two parallel north-south chains, separated from their nearest neighbors by a hundred miles on average. Swells generated by distant storms near Alaska, Antarctica, California, and Indonesia travel thousands of miles to these low-lying spits of sand. When they hit, part of their energy is reflected back out to sea in arcs, like sound waves emanating from a speaker; another part curls around the atoll or island and creates a confused chop in its lee. Wave piloting is the art of reading—by feel and by sight—these and other patterns. Detecting the minute differences in what to an untutored eye looks no more meaningful than a washing-machine cycle allows a *ri-meto*, a person of the sea in Marshallese, to determine where the nearest solid ground is —and how far off it lies—long before it is visible.

In the 16th century, Ferdinand Magellan, searching for a new route to the nutmeg and cloves of the Spice Islands, sailed through the Pacific Ocean and named it "the peaceful sea" before he was stabbed to death in the Philippines. Only 18 of his 270 men survived the trip. When subsequent explorers, despite similar travails, managed to make landfall on the countless islands sprinkled across this expanse, they were surprised to find inhabitants with nary a galleon, compass, or chart. God had created them there, the explorers hypothesized, or perhaps the islands were the remains of a sunken continent. As late as the 1960s, Western scholars still insisted that indigenous methods of navigating by stars, sun, wind, and waves were not nearly accurate enough, nor indigenous boats seaworthy enough, to have reached these tiny habitats on purpose.

Archaeological and DNA evidence (and replica voyages) have since proved that the Pacific islands were settled intentionally—by descendants of the first humans to venture out of sight of land, beginning some 60,000 years ago, from Southeast Asia to the Solomon Islands. They reached the Marshall Islands about 2,000 years ago. The geography of the archipelago that made wave piloting possible also made it indispensable as the sole means of collecting food, trading goods, waging war, and locating unrelated sexual partners. Chiefs threatened to kill anyone who revealed navigational knowledge without permission. In order to become a *ri-meto*, you had to be trained by a *ri-meto* and then pass a voyaging test, devised by your chief, on the first try. As colonizers from Europe introduced easier ways to get around, the training of *ri-metos* declined and became restricted primarily to an outlying atoll called Rongelap, where a shallow circular reef, set between ocean and lagoon, became the site of a small wave-piloting school.

In 1954 an American hydrogen-bomb test less than a hundred miles away rendered Rongelap uninhabitable. Over the next decades no new *ri-metos* were recognized; when the last well-known one died, in 2003, he left a 55-year-old cargo-ship captain named Korent Joel, who had trained at Rongelap as a boy, the effective custodian of their people's navigational secrets. Because of the radioactive fallout, Joel had not taken his voyaging test and thus was not a true *ri-meto*. But fearing that the knowledge might die with him, he asked for and received historic dispensation from his chief to train his younger cousin, Alson Kelen, as a wave pilot.

Now, in the lurching cabin of the power boat, Genz worried about whether Kelen knew what he was doing. Because Kelen was not a *ri-meto*, social mores forced him to insist that he was not navigating but *kajjidede*, or guessing. The sea was so rough tonight, Genz thought, that even for Joel, picking out a route would be like trying to hear a whisper in a gale. A voyage with this level of navigational difficulty had never been undertaken by anyone who was not a *ri-meto* or taking his test to become one. Genz steeled himself for the possibility that he might have to intervene for safety's sake, even if this was the best chance that he and his colleagues might ever get to unravel the scientific mysteries of wave piloting—and Kelen's best chance to rally support for preserving it. Organizing this trip had cost $72,000 in research grants, a fortune in the Marshalls.

The radio crackled. "*Jebro, Jebro,* this is *Jitdam,*" Kelen said. "Do you copy? Over."

Genz swallowed. The cabin's confines, together with the boat's diesel odors, did nothing to allay his motion sickness. "Copy that," he said. "Do you know where you are?"

*

Though mankind has managed to navigate itself across the globe and into outer space, it has done so in defiance of our innate way-finding capacities (not to mention survival instincts), which are still those of forest-dwelling homebodies. Other species use far more sophisticated cognitive methods to orient themselves. Dung beetles follow the Milky Way; the *Cataglyphis* desert ant dead-reckons by counting its paces; monarch butterflies, on their thousand-mile, multigenerational flight from Mexico to the Rocky Mountains, calculate due north using the position of the sun, which requires accounting for the time of day, the day of the year, and latitude; honeybees, newts, spiny lobsters, sea turtles, and many others read magnetic fields. Last year the fact of a "magnetic sense" was confirmed when Russian scientists put reed warblers in a cage that simulated different magnetic locations and found that the warblers always tried to fly "home" relative to whatever the programmed coordinates were. Precisely how the warblers detected these coordinates remains unclear. As does, for another example, the uncanny capacity of godwits to hatch from their eggs in Alaska and alone, without ever stopping, take off for French Polynesia. Clearly they and other long-distance migrants inherit a mental map and the ability to constantly recalibrate it. What it looks like in their mind's eye, however, and how it is maintained day and night, across thousands of miles, is still a mystery.

Efforts to scientifically deduce the neurological underpinnings of navigational abilities in humans and other species arguably began in 1948. An American psychologist named Edward Tolman made the heretical assertion that rats, until then regarded as mere slaves to behavioral reinforcement or punishment, create "cognitive maps" of their habitat. Tolman let rats accustom themselves to a maze with food at the end; then, leaving the food in the same spot, he rearranged the walls to introduce shortcuts—which the rodents took to reach the reward. This suggested that their sampling of various routes had given them a picture of the maze as a whole. Tolman hypothesized that humans have cognitive maps too, and that they are not just spatial but social. "Broad cognitive maps," he posited, lead to empathy, while narrow ones lead to "dangerous hates of outsiders," ranging from "discrimination against minorities to world conflagrations." Indeed, anthropologists today, especially those working in the Western Pacific, are increasingly aware of the potential ways in which people's physical environment—and how they habitually move through it—may shape their social relationships and how those ties may in turn influence their orienteering.

The cognitive map is now understood to have its own physical location, as a collection of electrochemical firings in the brain. In 1971, John O'Keefe, a neuroscientist at University College London, and a colleague reported that it had been pinpointed in the limbic system, an evolutionarily primitive region largely responsible for our emotional lives—specifically, within the hippocampus, an area where memories form. When O'Keefe implanted electrodes in rats' hippocampuses and measured their neural activity as they traveled through a maze, he detected "place cells" firing to mark their positions. In 1984, James B. Ranck Jr., a physiologist at the State University of New York, identified cells in an adjacent part of the brain that became active depending on the direction a rat's head was pointing—here was a kind of compass. And in 2005, building on these discoveries, Edvard and May-Britt Moser, neuroscientists at the Kavli Institute for Systems Neuroscience in Norway, found that our brains overlay our surroundings with a pattern of triangles. Anytime we reach an apex of one, a "grid cell" in an area of the brain in constant dialogue with the hippocampus delineates our position relative to the rest of the matrix. In 2014, O'Keefe and the Mosers shared a Nobel Prize for their discoveries of this "inner GPS" that constantly and subconsciously computes location.

The discovery that human orientation takes place in memory's seat—researchers have long known that damage to the hippocampus can cause amnesia—has raised the tantalizing prospect of a link between the two. In the late 1990s, Eleanor Maguire, a neuroscientist at University College London, began studying London taxi drivers, who must memorize the city's complex layout to obtain a license. Eventually she showed that when cabbies frequently access and revise their cognitive map, parts of their hippocampuses become larger; when they retire, those parts shrink. By contrast, following a sequence of directional instructions, as we do when using GPS, does not activate the hippocampus at all, according to work done by Veronique Bohbot, a cognitive neuroscientist at McGill University.

Bohbot and others are now trying to determine what effect, if any, the repeated bypassing of this region of the brain might be having on us. The hippocampus is one of the first areas disrupted by Alzheimer's disease, an early symptom of which is disorientation; shrinkage in the hippocampus and neighboring regions appears to increase the risk of depression, schizophrenia, and post-traumatic stress disorder. On the other hand, the taxi drivers who exercised their hippocampuses so much that parts of them changed size were worse at other memory tasks—and their performance on those improved after they retired. Few of us spend all day every day navigating, however, as cabbies do, and Maguire doubts that our GPS use is extreme enough to transform our gray matter.

What seems clear is that our ability to navigate is inextricably tied not just to our ability to remember the past but also to learning, decision-making, imagining, and planning for the future. And though our sense of direction often feels innate, it may develop —and perhaps be modified—in a region of the brain called the retrosplenial cortex, next to the hippocampus, which becomes active when we investigate and judge the permanence of landmarks. In 2012, Maguire and coauthors published their finding that an accurate understanding of whether a landmark is likely to stay put separates good navigators from poor ones, who are as apt to take cues from an idling delivery truck as a church steeple. The retrosplenial cortex passes our decisions about the stability of objects to the hippocampus, where their influence on way-finding intersects with other basic cognitive skills that, like memory, are as crucial to identity as to survival.

Recently Maguire and colleagues proposed a new unified theory of the hippocampus, imagining it not as a repository for disparate memories and directions but as a constructor of scenes that incorporate both. (Try to recall a moment from your past or picture a future one without visualizing yourself in the physical space where that moment happens.) Edvard and May-Britt Moser have similarly hypothesized that our ability to time-travel mentally evolved directly from our ability to travel in the physical world, and that the mental processes that make navigation possible are also the ones that allow us to tell a story. "In the same way that an infinite number of paths can connect the origin and endpoint of a journey," Edvard Moser and another coauthor wrote in a 2013 paper, "a recalled story can be told in many ways, connecting the beginning and the end through innumerable variations."

Disorientation is always stressful, and before modern civilization it was often a death sentence. Sometimes it still is. But recent studies have shown that people who use GPS, when given a pen and paper, draw less precise maps of the areas they travel through and remember fewer details about the landmarks they pass; paradoxically, this seems to be because they make fewer mistakes getting to where they're going. Being lost—assuming, of course, that you are eventually found—has one obvious benefit: the chance to learn about the wider world and reframe your perspective. From that standpoint, the greatest threat posed by GPS might be that we never do not know exactly where we are.

Genz took his thumb off the radio receiver's talk button and waited for Kelen's reply. He could make out on deck John Huth, a Harvard physicist and member of the international team that discovered the Higgs boson particle, vomiting volubly off the port side. The last time Genz checked, Gerbrant van Vledder, an oceanographer at Delft University in the Netherlands, one of the world's foremost institutions for wave modeling, was huddled miserably behind the abandoned galley, where a lone cabbage thudded against the walls of the sink. Compounding their digestive distress, a booby, ignoring the limitations of its webbed feet, had crash-landed on the deck, barring the men's access to the head. Sometimes Genz felt that all his decade of research on wave piloting had taught him was that he could never hope to predict what might go wrong next.

Genz met Alson Kelen and Korent Joel in Majuro in 2005, when Genz was 28. A soft-spoken, freckled Wisconsinite and former Peace Corps volunteer who grew up sailing with his father, Genz was then studying for a doctorate in anthropology at the University of Hawaii. His adviser there, Ben Finney, was an anthropologist who helped lead the voyage of *Hokulea*, a replica Polynesian sailing canoe, from Hawaii to Tahiti and back in 1976; the success of the trip, which involved no modern instrumentation and was meant to prove the efficacy of indigenous ships and navigational methods, stirred a resurgence of native Hawaiian language, music, hula, and crafts. Joel and Kelen dreamed of a similar revival for Marshallese sailing—the only way, they figured, for wave piloting to endure—and contacted Finney for guidance. But Finney was nearing retirement, so he suggested that Genz go in his stead. With their chief's blessing, Joel and Kelen offered Genz rare access, with one provision: he would not learn wave piloting himself; he would simply document Kelen's training.

Joel immediately asked Genz to bring scientists to the Marshalls who could help Joel understand the mechanics of the waves he knew only by feel—especially one called *di lep*, or backbone, the foundation of wave piloting, which (in *ri-meto* lore) ran between atolls like a road. Joel's grandfather had taught him to feel the *di lep* at the Rongelap reef: he would lie on his back in a canoe, blindfolded, while the old man dragged him around the coral, letting him experience how it changed the movement of the waves.

But when Joel took Genz out in the Pacific on borrowed yachts and told him they were encountering the *di lep*, he couldn't feel it. Kelen said he couldn't either. When oceanographers from the University of Hawaii came to look for it, their equipment failed to detect it. The idea of a wave-road between islands, they told Genz, made no sense.

Privately Genz began to fear that the *di lep* was imaginary, that wave piloting was already extinct. On one research trip in 2006, when Korent Joel went below deck to take a nap, Genz changed the yacht's course. When Joel awoke, Genz kept him away from the GPS device, and to the relief of them both, Joel directed the boat toward land. Later he also passed his *ri-meto* test, judged by his chief, with Genz and Kelen crewing.

Worlds away, Huth, a worrier by nature, had become convinced that preserving mankind's ability to way-find without technology was not just an abstract mental exercise but also a matter of life and death. In 2003, while kayaking alone in Nantucket Sound, fog descended, and Huth—spring-loaded and boyish, with a near-photographic memory—found his way home using local landmarks, the wind, and the direction of the swells. Later he learned that two young undergraduates, out paddling in the same fog, had become disoriented and drowned. This prompted him to begin teaching a class on primitive navigation techniques. When Huth met Genz at an academic conference in 2012 and described the methodology of his search for the Higgs boson and dark energy—subtracting dominant wave signals from a field until a much subtler signal appears underneath—Genz told him about the *di lep*, and it captured Huth's imagination. If it was real, and if it really ran back and forth between islands, its behavior was unknown to physics and would require a supercomputer to model. That a person might be able to sense it bodily amid the cacophony generated by other ocean phenomena was astonishing.

Huth began creating possible *di lep* simulations in his free time and recruited van Vledder's help. Initially the most puzzling detail of Genz's translation of Joel's description was his claim that the *di lep* connected each atoll and island to all 33 others. That would yield 561 paths, far too many for even the most adept wave pilot to memorize. Most of what we know about ocean waves and currents —including what will happen to coastlines as climate change leads to higher sea levels (of special concern to the low-lying Netherlands and Marshall Islands)—comes from models that use global wind and bathymetry data to simulate what wave patterns probably look like at a given place and time. Our understanding of wave mechanics, on which those models are based, is wildly incomplete. To improve them, experts must constantly check their assumptions with measurements and observations. Perhaps, Huth and van Vledder thought, there were *di leps* in every ocean, invisible roads that no one was seeing because they didn't know to look.

Early last year Genz and Kelen, grants in hand, saw a chance to show Huth and van Vledder the *di lep*. Kelen is the director of Waan Aelōñ in Majel, or Canoes of the Marshall Islands, a non-profit organization that teaches students to build canoes using traditional methods and modern materials. If the students hurried, the first sailing canoe to be built in the Marshalls in decades—the *Jitdam Kapeel,* which can be roughly translated as "the sharing of knowledge"—could be ready by summer's sailing season. Kelen's goal is for his students to build, staff, and maintain a fleet that will transport goods and passengers between atolls and islets without using fossil fuels. Despite the expectation that the Marshalls will be one of the first countries to disappear beneath rising seas, Kelen envisions a renaissance of sailing: a means for his students to reclaim their heritage while creating jobs that don't contribute to their own destruction.

Huth and van Vledder bought plane tickets to Majuro while Genz and Kelen made arrangements for the journey. At the last minute Joel's leg became infected, and Kelen offered to pilot in his place. The scientists embraced this new plan: talking with Joel before and after, they figured, would be almost as useful as having him onboard.

Soon after arriving they visited him at home, where he was confined to bed, and eagerly showed him their maps and simulations while posing detailed queries about various properties of the *di lep*. Although this was the scientific investigation Joel had been pushing for, he seemed reluctant to respond. He asked Huth and van Vledder if they believed in the *di lep;* they still weren't sure, they replied. Holding a rudimentary map that Huth had made of wave frequencies between Majuro and Aur, the captain traced a shaded region with his finger. "*Di lep* here," he said.

The next afternoon Kelen and his five-man crew set out for Aur. A breeze rattled the palms, blowing the *Jitdam* past a fleet of slumbering cargo ships anchored in the lagoon. The power boat *Jebro* puttered in pursuit. At the mouth of the opening between islets into the Pacific, the setting sun threw a flickering train on the water. "Now we get the truth," Huth cried, thrusting a sextant toward the sky. "The moment of reckoning!"

Twelve hours later Huth was seasick, bent over the deck rail, to which he had bound himself with a harness and tether. "If anyone said the *di lep* was subtle, they were wrong," he said, wiping his mouth. Nevertheless, he was doggedly recording on the hour the boat's GPS coordinates, the wind speed and direction, and his observations of the waves in a waterproof notebook. This data would allow him to map the journey with wind and wave details at each coordinate; van Vledder could later add wind data collected by satellite and local bathymetry, using programs written at Delft, to create a computer model of the seas they were currently in.

In the cabin Genz heard Kelen's voice on the radio again. Kelen could see the lights of the *Jebro* behind him, he said, and he thought they were about 10 miles east of Aur. Because they were approaching its reef too fast, his plan was to overshoot it, then look for it to his west after sunrise. Genz glanced at the boat's GPS device and realized that Kelen, over the last decade, might have learned more than he had ever let on. He wanted to shout congratulations.

"Copy that," he said instead.

The sky grew lighter, revealing more sky, a flock of seabirds fishing, and finally, far ahead, the canoe, battered but intact, struggling to head downwind. After getting a brief tow from the *Jebro,* it reached Aur under its own power. An empty beach came into view, then children running on it. "This is feeling like an adventurer," van Vledder said. "Coming to a new place, and people out to welcome you."

The entire village was waiting in a palm-frond-thatched pavilion, having been alerted by ham radio. A woman put leis around the necks of the sailors and scientists as they entered. The community had piled a long table with lobster, fish, breadfruit, plantains, and rice balls with coconut. The acting chief of the island made a speech. He said the local children had never seen a sailing canoe before. The islanders wanted to learn to build them again; they had only one motorboat, and gasoline there cost more per gallon than most of them made in a month of selling fish and handicrafts in Majuro.

Two mornings later Kelen stood outside a cinder-block schoolhouse on Aur that the chief had offered as a dormitory, looking up at an overcast sky and weighing again—as he had when he first met Genz—how much of his knowledge to share in order to keep it alive. Now in his late 40s and newly a grandfather, he had lived his early childhood on the atoll nearest Rongelap, Bikini, where the hydrogen bomb and dozens of other nuclear weapons were exploded. Later, as part of a program to test the effects of radiation on humans, American officials told the people from Bikini and Rongelap that their islands were safe to resettle, so they returned for several years. During this period Kelen's father taught him to sail in a traditional canoe made by Kelen's grandfather. When Kelen was 10, the Americans finally evacuated the islanders to Kili, an uninhabited island bedeviled on all sides by violent ocean swells too rough for the canoe, which rotted away.

Eventually Kelen's parents moved to Majuro, home to half of the nation's 50,000 citizens—an urban hub compared with the outer islands. They sent Kelen, a top student, to boarding school in Honolulu. There, when he was 19, he went with his class down to the docks to watch the world-famous *Hokulea* return from a trip to New Zealand. Later he came back to Majuro as a young man and dedicated himself to the preservation of fading skills, like weaving and canoe building. But he felt tremendous ambivalence about what gaining resources to preserve his culture, or any native

culture, seemed to require: allowing outsiders, whether academics or reporters, to commodify it. Secrecy and hands-on training
are integral to the tradition of wave piloting; explaining the *di lep*
would disrupt those features of it even while immortalizing it in
books and journals, perhaps inspiring more Marshallese children
to become *ri-metos*.

The tide was on its way out as the sailors and scientists began to
load up for the 70-mile journey back to Majuro. The villagers sang
again and prayed for their safe return. They laid another feast
and stocked the canoe with provisions, packed in woven pandanus
baskets, and handicrafts, including a toy sailing canoe, a perfect
imitation, small and light as a bird. Until now, because his crew
and canoe were untested, Kelen had deemed it unsafe to have any
passengers aboard the *Jitdam*. One more person could fit, however,
and he invited me onboard.

"*Youp, youp,*" called Binton Daniel, the master builder who had
supervised the construction of the *Jitdam*, and the sail shot up. The
sailors waved in overhead arcs at the people on the beach. The
people waved back. Gradually the sound of swells rushing against
the coral rim of the lagoon grew louder. With a *thunk*, the bottom
of the canoe hit the top of the reef and slid across, and we were
out in open water.

Daniel eased the mainsheet and let the boom swing out. The first
mate, Jason Ralpho, a stern-looking man in gray socks who worked
with Kelen at the Ports Authority, and Ejnar Aerok, a plump, professional karaoke singer, secured the line to a cleat. The youngest,
Elmi Juonran, lifted a lid off one of two hatches and, muttering,
disappeared to boil water for ramen in a big silver teakettle. "He
says he's the only one who knows the password to these doors,"
Kelen said. Juonran's cousin, Sear Helios, named for the department store his parents visited on a trip to Honolulu, steered from
the stern of the canoe with a 50-pound wooden paddle.

Kelen leaned back against the mast and looked at the front of
the outrigger float and the back, estimating our speed. He checked
his wristwatch. The wind was coming from the northeast, and the
current, he said, would take us farther east that night. Ostensi-
bly he was dead-reckoning—to do that you must know where you
started, where you're going, how fast you're moving, and in what
direction. Wave piloting, if Genz, Huth, and van Vledder are right,

is more precise; theoretically, a wave pilot dropped blindfolded into a boat in Marshallese waters could follow a set of seamarks —waves of a particular shape—alone to land.

"Majuro should be that way," Kelen said, pointing. "I'm closing my eyes and looking at the wind. This is a very short distance. Again, I'm only a student. I'm entitled to a few mistakes."

Swells glided, smooth and gentle, beneath us. Sunset cracked yolk on a puffy lavender sky. The horizon appeared infinite and also very near, as if we had fallen into a mixing bowl. Around us the crew faded into shadow. Ralpho lit a cigarette, and its tip burned orange in the dark. The sail luffed.

"This is kind of scary smooth," Kelen said. "Does it feel like we're moving anywhere? That's not good. We have to move or we'll drift away from the islands." Yet he didn't sound worried. We lay back. The sky was foggy with stars.

As a young man, Kelen said, he spent some time on the West Coast, picking strawberries in Oregon, working in a turkey plant, then driving a Rent Town USA truck up and down Highway 101. He described long days of sweet berries, of cutting the necks of birds, of truck-stop sloppy Joes and giant cups of coffee. We lost sight of the *Jebro* and missed three call-ins. Kelen could still remember fishing as a child on Bikini, its long white beaches. In his memory, everyone there was happy. Periodically a government ship brought provisions, and men in white lab coats tested him and the other islanders with a huge machine. When the ship came to take them away for good, Kelen thought they were going for a ride.

Aerok began to sing in a high, lonely tenor. Ralpho added baritone harmony. "It's kind of like a country-music song," Kelen said. "'I see you as beautiful as a sunset, and I cry when I leave the beach that you stand on.' It's kind of like a sailors' leaving-home song. It's a song when you start singing it, everyone knows it."

I closed my eyes. The sounds of the canoe—creaking, sloshing, rippling—traced its shape like fingers moving over a face in the dark.

I awoke to Aerok and Juonran singing about Majuro, another sad song. The sky spilled radiance onto the water. Beside me, Kelen was awake too. "Every time I look up at heaven, I wonder, how many Earths are out there?" he said. "How many planets like ours? There's millions of galaxies. There must be something."

We saw one star drop, then another. "Every time I see a fall-ing star, I make a wish and I don't tell nobody," he said. "I don't believe very many things, but this is something that makes me feel good, even if it isn't true."

By 9:30 the next morning the sun was high and the sailors had grown quiet. Kelen rested his shoulder against the mast, peering into the distance. If we didn't see Majuro by 10, he said, the cur-rent had pushed us too far to the west. At 9:50 Juonran pointed, and everyone else followed his finger to the faintest of tints on the horizon. Kelen swatted him on the butt. The sailors laughed.

"Another good guess," Kelen said to me.

All maps are but representations of reality: they render the physi-cal world in symbols and highlight important relationships—the proximity of one subway stop to another, say—that are invisible to the naked eye. If storytelling, the way we structure and make meaning from the events of our lives, arose from navigating, so too is the practice of navigation inherently bound up with storytelling, in all its subjectivity.

"When I was young, we had canoes," Kelen told me one after-noon on Aur. "We didn't have TVs. In evening time my father would open his arm, like this, and say, *Lie there*"—he tapped the inside of his elbow—"and he would tell me the legends of sail-ing. Some people have those heroes, like Superman, and they're picturing they are Superman. When my dad talked about sailing, I was on that canoe."

To teach way-finding, the Marshallese use stick charts, wood frames crosshatched like dream catchers to represent swells com-ing from four cardinal directions, with shells woven in to symbol-ize the position of the atolls. These meant nothing to the first Eu-ropean explorers to see them, just as Mercator projections meant nothing to the Marshallese. Even today local schoolchildren visit-ing the historical museum in Majuro are sometimes baffled when they're told that the blue-and-green pictures on the walls are pic-tures of where they are.

If "where" is both subjective and physical, what do you need to know, precisely, to figure out where you are? From the moment our nomad ancestors wandered out of Africa until a few decades ago, locating yourself required interacting in some way with the environment: following the stars or a migrating herd of wildebeests, even reading a compass or a street sign. Then, in the time it took to transition from rotary phones to smartphones, we became the first unnatural long-distance migrants, followers of step-by-step instructions that obviated the need to look around at all. Over the last several years, organizations like the United States military and the Federal Aviation Administration have expressed concern about their overwhelming reliance on GPS and the possibility that the network's satellite signals could be sabotaged by an enemy or disabled by a strong solar flare. The United States Naval Academy has once again begun training midshipmen in how to take their position from the stars with a sextant.

As researchers urgently explore what GPS is doing to our minds, wave piloting—a technique that seems to involve the subtlest environmental cues a person can detect—is slipping, virtually unnoticed, from human consciousness. Even if Huth and van Vledder could figure out how it worked, they admitted, it didn't mean they could feel it or teach others how to do so.

Back on Majuro, they spent several days typing notes and crunching data, barely emerging from their rooms. Huth created a preliminary map of the route and approximate wind and sea conditions to show Korent Joel to see if he could identify a pattern that might be the *di lep*. But when they arrived at his home again, they learned that he had checked into the hospital the previous afternoon. Several weeks later he was flown to Honolulu, where surgeons determined that his leg was gangrenous and amputated it below the knee. In his absence, Kelen and Genz helped Huth and van Vledder quiz Joel's Rongelapese uncle for stray clues to *di lep*'s features, but learned nothing they recognized as epiphanies.

Until November, when van Vledder visited Cambridge, Massachusetts, where he and Huth sequestered themselves in Huth's office. As they mapped the coordinates Huth had recorded atop van Vledder's model of sea conditions, they found that the path they had taken was exactly perpendicular to a dominant eastern swell flowing between Majuro and Aur. And at places where the swell, influenced by the surrounding atolls, turned slightly northeast or southeast, the path bent to match. It was a curve. Everyone had assumed that a wave called "backbone" would look like one. "But nobody said the *di lep* is a straight line," van Vledder said.

What if, they conjectured, the "road" isn't a single wave reflecting back and forth between every possible combination of atolls and islands; what if it is the path you take if you keep your vessel at 90 degrees to the strongest swell flowing between neighboring bodies of land? Position your broadside correctly, smack in the *di lep*'s path, and your hull would rock symmetrically, side to side—in a manner that would turn a loose cabbage into a pendulum and teach an anthropologist, a physicist, and an oceanographer a hard lesson about the human gastrointestinal system's adaptation to life at sea. In other words, it was as Joel's uncle had, it turned out, told them: the *di lep* feels like *pidodo*, diarrhea. We might have been riding it all along.

NICOLA TWILLEY

The Billion-Year Wave

FROM *The New Yorker*

JUST OVER A billion years ago, many millions of galaxies from
here, a pair of black holes collided. They had been circling each
other for aeons in a sort of mating dance, gathering pace with
each orbit, hurtling closer and closer. By the time they were a few
hundred miles apart, they were whipping around at nearly the
speed of light, releasing great shudders of gravitational energy.
Space and time became distorted, like water at a rolling boil. In
the fraction of a second that it took for the black holes to finally
merge, they radiated a hundred times more energy than all the
stars in the universe combined. They formed a new black hole, 62
times as heavy as our sun and almost as wide across as the state of
Maine. As it smoothed itself out, assuming the shape of a slightly
flattened sphere, a few last quivers of energy escaped. Then space
and time became silent again.

The waves rippled outward in every direction, weakening as
they went. On Earth, dinosaurs arose, evolved, and went extinct.
The waves kept going. About 50,000 years ago they entered our
own Milky Way galaxy, just as *Homo sapiens* were beginning to replace our Neanderthal cousins as the planet's dominant species
of ape. A hundred years ago, Albert Einstein, one of the more
advanced members of the species, predicted the waves' existence,
inspiring decades of speculation and fruitless searching. Twenty-
two years ago, construction began on an enormous detector, the
Laser Interferometer Gravitational-Wave Observatory (LIGO).
Then, on September 14, 2015, at just before 11 in the morning,
Central European Time, the waves reached Earth. Marco Drago,

a 32-year-old Italian postdoctoral student and a member of the
LIGO Scientific Collaboration, was the first person to notice them.
He was sitting in front of his computer at the Albert Einstein Institute, in Hannover, Germany, viewing the LIGO data remotely. The
waves appeared on his screen as a compressed squiggle, but the
most exquisite ears in the universe, attuned to vibrations of less
than a trillionth of an inch, would have heard what astronomers
call a chirp—a faint whooping from low to high. This morning,
in a press conference in Washington, D.C., the LIGO team announced that the signal constitutes the first direct observation of
gravitational waves.

When Drago saw the signal, he was stunned. "It was difficult
to understand what to do," he told me. He informed a colleague,
who had the presence of mind to call the LIGO operations room,
in Livingston, Louisiana. Word began to circulate among the thousand or so scientists involved in the project. In California, David
Reitze, the executive director of the LIGO Laboratory, saw his
daughter off to school and went to his office at Caltech, where he
was greeted by a barrage of messages. "I don't remember exactly
what I said," he told me. "It was along these lines: 'Holy shit, what
is this?'" Vicky Kalogera, a professor of physics and astronomy at
Northwestern University, was in meetings all day and didn't hear
the news until dinnertime. "My husband asked me to set the table," she said. "I was completely ignoring him, skimming through
all these weird emails and thinking, 'What is going on?'" Rainer
Weiss, the 83-year-old physicist who first suggested building LIGO,
in 1972, was on vacation in Maine. He logged on, saw the signal,
and yelled "My God!" loudly enough that his wife and adult son
came running.

The collaborators began the arduous process of double-, triple-,
and quadruple-checking their data. "We're saying that we made a
measurement that is about a thousandth the diameter of a proton,
that tells us about two black holes that merged over a billion years
ago," Reitze said. "That is a pretty extraordinary claim and it needs
extraordinary evidence." In the meantime the LIGO scientists
were sworn to absolute secrecy. As rumors of the finding spread,
from late September through this week, media excitement spiked;
there were rumblings about a Nobel Prize. But the collaborators
gave anyone who asked about it an abbreviated version of the truth

—that they were still analyzing data and had nothing to announce. Kalogera hadn't even told her husband.

LIGO consists of two facilities separated by nearly 1,900 miles— about a three-and-a-half-hour flight on a passenger jet, but a journey of less than ten thousandths of a second for a gravitational wave. The detector in Livingston, Louisiana, sits on swampland east of Baton Rouge, surrounded by a commercial pine forest; the one in Hanford, Washington, is on the southwestern edge of the most contaminated nuclear site in the United States, amid desert sagebrush, tumbleweed, and decommissioned reactors. At both locations a pair of concrete pipes some 12 feet tall stretch at right angles into the distance, so that from high above the facilities resemble carpenter's squares. The pipes are so long—nearly two and a half miles—that they have to be raised from the ground by a yard at each end, to keep them lying flat as Earth curves beneath them.

LIGO is part of a larger effort to explore one of the more elusive implications of Einstein's general theory of relativity. The theory, put simply, states that space and time curve in the presence of mass, and that this curvature produces the effect known as gravity. When two black holes orbit each other, they stretch and squeeze spacetime like children running in circles on a trampoline, creating vibrations that travel to the very edge; these vibrations are gravitational waves. They pass through us all the time, from sources across the universe, but because gravity is so much weaker than the other fundamental forces of nature—electromagnetism, for instance, or the interactions that bind an atom together—we never sense them. Einstein thought it highly unlikely that they would ever be detected. He twice declared them nonexistent, reversing and then re-reversing his own prediction. A skeptical contemporary noted that the waves seemed to "propagate at the speed of thought."

Nearly five decades passed before someone set about building an instrument to detect gravitational waves. The first person to try was an engineering professor at the University of Maryland, College Park, named Joe Weber. He called his device the resonant bar antenna. Weber believed that an aluminum cylinder could be made to work like a bell, amplifying the feeble strike of a gravitational wave. When a wave hit the cylinder, it would vibrate very slightly, and sensors around its circumference would translate the ringing into an electrical signal. To make sure he wasn't detecting the vibrations of passing trucks or minor earthquakes, Weber developed several safeguards: he suspended his bars in a vacuum, and he ran two of them at a time, in separate locations—one on the campus of the University of Maryland and one at Argonne National Laboratory, near Chicago. If both bars rang in the same way within a fraction of a second of each other, he concluded, the cause might be a gravitational wave.

In June of 1969, Weber announced that his bars had registered something. Physicists and the media were thrilled; the *Times* reported that "a new chapter in man's observation of the universe has been opened." Soon Weber started reporting signals on a daily basis. But doubt spread as other laboratories built bars that failed to match his results. By 1974 many had concluded that Weber was mistaken. (He continued to claim new detections until his death, in 2000.)

Weber's legacy shaped the field that he established. It created a poisonous perception that gravitational-wave hunters, as Weiss put it, are "all liars and not careful, and God knows what." That perception was reinforced in 2014, when scientists at BICEP2, a telescope near the South Pole, detected what seemed to be gravitational radiation left over from the Big Bang; the signal was real, but it turned out to be a product of cosmic dust. Weber also left behind a group of researchers who were motivated by their inability to reproduce his results. Weiss, frustrated by the difficulty of teaching Weber's work to his undergraduates at the Massachusetts Institute of Technology, began designing what would become LIGO. "I couldn't understand what Weber was up to," he said in an oral history conducted by Caltech in 2000. "I didn't think it was right. So I decided I would go at it myself."

In the search for gravitational waves, "most of the action takes place on the phone," Fred Raab, the head of LIGO's Hanford site, told me. There are weekly meetings to discuss data and fortnightly meetings to discuss coordination between the two detectors, with collaborators in Australia, India, Germany, the United Kingdom, and elsewhere. "When these people wake up in the middle of the night dreaming, they're dreaming about the detector," Raab said.

"That's how intimate they have to be with it," he explained, to be able to make the fantastically complex instrument that Weiss conceived actually work.

Weiss's detection method was altogether different from Weber's. His first insight was to make the observatory *L*-shaped. Picture two people lying on the floor, their heads touching, their bodies forming a right angle. When a gravitational wave passes through them, one person will grow taller while the other shrinks; a moment later the opposite will happen. As the wave expands spacetime in one direction, it necessarily compresses it in the other. Weiss's instrument would gauge the difference between these two fluctuating lengths, and it would do so on a gigantic scale, using miles of steel tubing. "I wasn't going to be detecting anything on my tabletop," he said.

To achieve the necessary precision of measurement, Weiss suggested using light as a ruler. He imagined putting a laser in the crook of the *L*. It would send a beam down the length of each tube, which a mirror at the other end would reflect back. The speed of light in a vacuum is constant, so as long as the tubes were cleared of air and other particles the beams would recombine at the crook in synchrony—unless a gravitational wave happened to pass through. In that case, the distance between the mirrors and the laser would change slightly. Since one beam would now be covering a shorter distance than its twin, they would no longer be in lockstep by the time they got back. The greater the mismatch, the stronger the wave. Such an instrument would need to be thousands of times more sensitive than any previous device, and it would require delicate tuning in order to extract a signal of vanishing weakness from the planet's omnipresent din.

Weiss wrote up his design in the spring of 1972, as part of his laboratory's quarterly progress report. The article never appeared in a scientific journal—it was an idea, not an experiment—but according to Kip Thorne, an emeritus professor at Caltech who is perhaps best known for his work on the movie *Interstellar*, "it is one of the greatest papers ever written." Thorne doesn't recall reading Weiss's report until later. "If I had read it, I had certainly not understood it," he said. Indeed, Thorne's landmark textbook on gravitational theory, coauthored with Charles Misner and John Wheeler and first published in 1973, contained a student exercise designed to demonstrate the impracticability of measuring gravitational waves with lasers. "I turned around on that pretty quickly," he told me.

Thorne's conversion occurred in a hotel room in Washington, D.C., in 1975. Weiss had invited him to speak to a panel of NASA scientists. The evening before the meeting, the two men got to talking. "I don't remember how it happened, but we shared the hotel room that night," Weiss said. They sat at a tiny table, filling sheet after sheet of paper with sketches and equations. Thorne, who was raised Mormon, drank Dr Pepper; Weiss smoked a corncob pipe stuffed with Three Nuns tobacco. "There are not that many people in the world that you can talk to like that, where both of you have been thinking about the same thing for years," Weiss said. By the time Thorne got back to his own room, the sky was turning pink.

At MIT, Weiss had begun assembling a small prototype detector with five-foot arms. But he had trouble getting support from departmental administrators, and many of his colleagues were also skeptical. One of them, an influential astrophysicist and relativity expert named Phillip Morrison, was firmly of the opinion that black holes did not exist—a viewpoint that many of his contemporaries shared, given the paucity of observational data. Since black holes were some of the only cosmic phenomena that could theoretically emit gravitational waves of significant size, Morrison believed that Weiss's instrument had nothing to find. Thorne had more success: by 1981 there was a prototype underway at Caltech, with arms 131 feet long. A Scottish physicist named Ronald Drever oversaw its construction, improving on Weiss's design in the process.

In 1990, after years of studies, reports, presentations, and committee meetings, Weiss, Thorne, and Drever persuaded the National Science Foundation to fund the construction of LIGO. The project would cost $272 million, more than any NSF-backed experiment before or since. "That started a huge fight," Weiss said. "The astronomers were dead set against it, because they thought it was going to be the biggest waste of money that ever happened." Many scientists were concerned that LIGO would sap money from other research. Rich Isaacson, a program officer at the NSF at the time, was instrumental in getting the observatory off the ground.

"He and the National Science Foundation stuck with us and took this enormous risk," Weiss said.

"It never should have been built," Isaacson told me. "It was a couple of maniacs running around, with no signal ever having been discovered, talking about pushing vacuum technology *and* laser technology *and* materials technology *and* seismic isolation and feedback systems orders of magnitude beyond the current state of the art, using materials that hadn't been invented yet." But Isaacson had written his PhD thesis on gravitational radiation, and he was a firm believer in LIGO's theoretical underpinnings. "I was a mole for the gravitational-wave community inside the NSF," he said.

In their proposal, the LIGO team warned that their initial design was unlikely to detect anything. Nonetheless, they argued, an imperfect observatory had to be built in order to understand how to make a better one. "There was every reason to imagine this was going to fail," Isaacson said. He persuaded the NSF that even if no signal was registered during the first phase, the advances in precision measurement that came out of it would likely be worth the investment. Ground was broken in early 1994.

It took years to make the most sensitive instrument in history insensitive to everything that is not a gravitational wave. Emptying the tubes of air demanded 40 days of pumping. The result was one of the purest vacuums ever created on Earth, a trillionth as dense as the atmosphere at sea level. Still, the sources of interference were almost beyond reckoning—the motion of the wind in Hanford or of the ocean in Livingston; imperfections in the laser light as a result of fluctuations in the power grid; the jittering of individual atoms within the mirrors; distant lightning storms. All can obscure or be mistaken for a gravitational wave, and each source had to be eliminated or controlled for. One of LIGO's systems responds to minuscule seismic tremors by activating a damping system that pushes on the mirrors with exactly the right counterforce to keep them steady; another monitors for disruptive sounds from passing cars, airplanes, or wolves.

"There are ten thousand other tiny things, and I really mean ten thousand," Weiss said. "And every single one needs to be working correctly so that nothing interferes with the signal." When his

colleagues make adjustments to the observatory's interior components, they must set up a portable clean room, sterilize their tools,
and don what they call bunny suits—full-body protective gear—
lest a skin cell or a particle of dust accidentally settles on the spar-
kling optical hardware.

The first iteration of the observatory—Initial LIGO, as the team
now calls it—was up and running in 2001. During the next nine
years the scientists measured and refined their instruments' performance and improved their data-analysis algorithms. In the meantime they used the prototype at Caltech and a facility in Germany
to develop ever more sensitive mirror, laser, and seismic-isolation
technology. In 2010 the detectors were taken offline for a five-year,
$250 million upgrade. They are now so well shielded that when
the facilities manager at the Hanford site revs his Harley next to
the control room, the scientist monitoring the gravitational-wave
channel sees nothing. (A test of this scenario is memorialized in
the logbook as "Bubba Roars Off on a Motor Cycle.") The observatory's second iteration, Advanced LIGO, should eventually be capable of surveying a volume of space that is more than a thousand
times greater than its predecessor's.

Some of the most painstaking work took place on the mirrors,
which, Reitze said, are the best in the world "by far." Each is a little
more than a foot wide, weighs nearly 90 pounds, and is polished to
within a hundred-millionth of an inch of a perfect sphere. (They
cost almost half a million dollars apiece.) At first the mirrors were
suspended from loops of steel wire. For the upgrade they were
attached instead to a system of pendulums, which insulated them
even further from seismic tremors. They dangle from fibers of
fused silica—glass, basically—which, although strong enough to
bear the weight of the mirrors, shatter at the slightest provocation.
"We did have one incident where a screw fell and pinged one, and
it just went *poof,*" Anamaria Effler, a former operations specialist
at the Hanford site, told me. The advantage of the fibers is their
purity, according to Jim Hough, of the University of Glasgow. "You
know how, when you flick a whiskey glass, it will ring beautifully?"
he asked. "Fused silica is even better than a whiskey glass—it is like
plucking a string on a violin." The note is so thin that it is possible
for LIGO's signal-processing software to screen it out—another
source of interference eliminated.

Preparing Advanced LIGO took longer than expected, so the

new and improved instrument's start date was pushed back a few days, to September 18, 2015. Weiss was called in from Boston a week prior to try to track down the source of some radio-frequency interference. "I get there and I was horrified," he said. "It was everywhere." He recommended a weeklong program of repairs to address the issue, but the project's directors refused to delay the start of the first observing run any longer. "Thank God they didn't let me do it," Weiss said. "I would have had the whole goddamn thing offline when the signal came in."

On Sunday, September 13, Effler spent the day at the Livingston site with a colleague, finishing a battery of last-minute tests. "We yelled, we vibrated things with shakers, we tapped on things, we introduced magnetic radiation, we did all kinds of things," she said. "And of course everything was taking longer than it was supposed to." At four in the morning, with one test still left to do—a simulation of a truck driver hitting his brakes nearby—they decided to pack it in. They drove home, leaving the instrument to gather data in peace. The signal arrived not long after, at 4:50 a.m. local time, passing through the two detectors within seven milliseconds of each other. It was four days before the start of Advanced LIGO's first official run.

The fact that gravitational waves were detected so early prompted confusion and disbelief. "I had told everyone that we wouldn't see anything until 2017 or 2018," Reitze said. Janna Levin, a professor of astrophysics at Barnard College and Columbia University, who is not a member of the LIGO Scientific Collaboration, was equally surprised. "When the rumors started, I was like, Come on!" she said. "They only just got it locked!" The signal, moreover, was almost too perfect. "Most of us thought that when we ever saw such a thing, it would be something that you would need many, many computers and calculations to drag out of the noise," Weiss said. Many of his colleagues assumed that the signal was some kind of test.

The LIGO team includes a small group of people whose job is to create blind injections—bogus evidence of a gravitational wave —as a way of keeping the scientists on their toes. Although everyone knew who the four people in that group were, "we didn't know what, when, or whether," Gabriela González, the collaboration's spokeswoman, said. During Initial LIGO's final run, in 2010, the detectors picked up what appeared to be a strong signal. The scientists analyzed it intensively for six months, concluding that it was a gravitational wave from somewhere in the constellation of Canis Major. Just before they submitted their results for publication, however, they learned that the signal was a fake.

This time through, the blind-injection group swore that they had nothing to do with the signal. Marco Drago thought that their denials might also be part of the test, but Reitze, himself a member of the quartet, had a different concern. "My worry was —and you can file this under the fact that we are just paranoid cautious about making a false claim—could somebody have done this maliciously?" he said. "Could somebody have somehow faked a signal in our detector that we didn't know about?" Reitze, Weiss, González, and a handful of others considered who, if anyone, was familiar enough with both the apparatus and the algorithms to have spoofed the system and covered his or her tracks. There were only four candidates, and none of them had a plausible motive. "We grilled those guys," Weiss said. "And no, they didn't do it." Ultimately, he said, "We accepted that the most economical explanation was that it really is a black-hole pair."

Subgroups within the LIGO Scientific Collaboration set about validating every aspect of the detection. They reviewed how the instruments had been calibrated, took their software code apart line by line, and compiled a list of possible environmental disturbances, from oscillations in the ionosphere to earthquakes in the Pacific Rim. ("There was a very large lightning strike in Africa at about the same time," Stan Whitcomb, LIGO's chief scientist, told me. "But our magnetometers showed that it didn't create enough of a disturbance to cause this event.") Eventually they confirmed that the detection met the statistical threshold of five sigma, the gold standard for declaring a discovery in physics. This meant that there was a probability of only one in 3.5 million that the signal was spotted by chance.

The September 14 detection, now officially known as GW150914, has already yielded a handful of significant astrophysical findings. To begin with, it represents the first observational evidence that black-hole pairs exist. Until now they had existed only in theory, since by definition they swallow all light in their vicinity, rendering themselves invisible to conventional telescopes. Gravitational waves are the only information known to be capable of escaping a black hole's crushing gravity.

The LIGO scientists have extracted an astonishing amount from the signal, including the masses of the black holes that produced it, their orbital speed, and the precise moment at which their surfaces touched. They are substantially heavier than expected, a surprise that, if confirmed by future observations, may help to explain how the mysterious supermassive black holes at the heart of many galaxies are formed. The team has also been able to quantify what is known as the ringdown—the three bursts of energy that the new, larger black hole gave off as it became spherical. "Seeing the ringdown is spectacular," Levin said. It offers confirmation of one of relativity theory's most important predictions about black holes—namely, that they radiate away imperfections in the form of gravitational waves after they coalesce.

The detection also proves that Einstein was right about yet another aspect of the physical universe. Although his theory deals with gravity, it has primarily been tested in our solar system, a place with a notably weak gravitational regime. "You think Earth's gravity is really something when you're climbing the stairs," Weiss said. "But as far as physics goes, it is a pipsqueak, infinitesimal, tiny little effect." Near a black hole, however, gravity becomes the strongest force in the universe, capable of tearing atoms apart. Einstein predicted as much in 1916, and the LIGO results suggest that his equations align almost perfectly with real-world observation. "How could he have ever known this?" Weiss asked. "I would love to present him with the data that I saw that morning, to see his face."

Since the September 14 detection, LIGO has continued to observe candidate signals, although none are quite as dramatic as the first event. "The reason we are making all this fuss is because of the big guy," Weiss said. "But we're very happy that there are other, smaller ones, because it says this is not some unique, crazy, cuckoo effect."

Virtually everything that is known about the universe has come to scientists by way of the electromagnetic spectrum. Four hundred years ago Galileo began exploring the realm of visible light with his telescope. Since then astronomers have pushed their instruments further. They have learned to see in radio waves and microwaves, in infrared and ultraviolet, in X-rays and gamma rays, revealing the birth of stars in the Carina Nebula and the eruption of geysers on Saturn's eighth moon, pinpointing the center of the Milky Way and the locations of Earth-like planets around us. But more than 95 percent of the universe remains imperceptible to traditional astronomy. Gravitational waves may not illuminate the so-called dark energy that is thought to make up the majority of that obscurity, but they will enable us to survey space and time as we never have before. "This is a completely new kind of telescope," Reitze said. "And that means we have an entirely new kind of astronomy to explore." If what we witnessed before was a silent movie, Levin said, gravitational waves turn our universe into a talkie.

As it happens, the particular frequencies of the waves that LIGO can detect fall within the range of human hearing, between about 35 and 250 hertz. The chirp was much too quiet to hear by the time it reached Earth, and LIGO was capable of capturing only two-tenths of a second of the black holes' multibillion-year merger, but with some minimal audio processing the event sounds like a glissando. "Use the back of your fingers, the nails, and just run them along the piano from the lowest A up to middle C, and you've got the whole signal," Weiss said.

Different celestial sources emit their own sorts of gravitational waves, which means that LIGO and its successors could end up hearing something like a cosmic orchestra. "The binary neutron stars are like the piccolos," Reitze said. Isolated spinning pulsars, he added, might make a monochromatic *ding* like a triangle, and black holes would fill in the string section, running from double bass on up, depending on their mass. LIGO, he said, will only ever be able to detect violins and violas; waves from supermassive black holes, like the one at the center of the Milky Way, will have to await future detectors, with different sensitivities.

Several such detectors are in the planning stages or under construction, including the Einstein Telescope, a European project whose underground arms will be more than twice the length of LIGO's, and a space-based constellation of three instruments called eLISA. (The European Space Agency, with support from NASA, launched a Pathfinder mission in December.) Other detectors are already up and running, including the BICEP2 telescope, which, despite its initial false alarm, may still detect the echoes of gravitational waves from even further back in the universe's history. Reitze's hope, he told me, is that the chirp will motivate more investment in the field.

Advanced LIGO's first observing run came to an end on January 12. Effler and the rest of the commissioning team have since begun another round of improvements. The observatory is inching toward its maximum sensitivity; within two or three years it may well register events on a daily basis, capturing more data in the process. It will come online again by late summer, listening even more closely to a celestial soundtrack that we have barely imagined. "We are opening up a window on the universe so radically different from all previous windows that we are pretty ignorant about what's going to come through," Thorne said. "There are just bound to be big surprises."

PART II

Changing Land and Resources

BECCA CUDMORE

The Case for Leaving
City Rats Alone

FROM *Nautilus*

KAYLEE BYERS CROUCHES in a patch of urban blackberries
early one morning this June to check a live trap in one of Vancouver's poorest areas, the V6A postal code. Her first catch of the day
is near a large blue dumpster on "Block 5," in front of a 20-some-
unit apartment complex above a thrift shop. Across the alley a
building is going up; between the two is an overgrown paper- and
wrapper-strewn lot. In the lot there are rats.

"Once we caught two in a single trap," she says, peering inside
the cage. She finds a new rat there and makes a note of it on her
clipboard; she'll be back for it, to take the animal to her nearby
van, which is parked near (according to Google Maps) an "un-
fussy" traditional Ethiopian restaurant. Once inside the van, the
rat will be put under anesthesia and will then be photographed,
brushed for fleas, tested for disease, fixed with an ear tag, and released back into V6A within 45 minutes.

Byers is a PhD student under veterinary pathologist Chelsea
Himsworth, a University of British Columbia School of Population
and Public Health assistant professor who has become a local science celebrity thanks to her Vancouver Rat Project. Himsworth
started the project as a way to address health concerns over the
city's exploding rat population — exploding anecdotally, that is, as
no one has counted it.

Prior to Himsworth's work, in fact, the sum total knowledge of
Canada's wild rats could be boiled down to a single study of 43 rats

living in a landfill in nearby Richmond in 1984. So six years ago she stocked an old minivan with syringes, needles, and gloves and live-trapped more than 700 of V6A's rats to sample their DNA and learn about the bacteria they carried.

Her research has made her reconsider the age-old labeling of rats as invaders that need to be completely fought back. They may instead be just as much a part of our city as sidewalks and lamp-posts. We would all be better off if, under most circumstances, we simply left them alone.

Rats thrive as a result of people. The great modern disruptions caused by urban development and human movement across the world have ferried them to new ecological niches. "Rats are real disturbance specialists," says biologist Ken Aplin, who has studied the rodents and their diseases for decades. "Very few wild animals have adapted so well to the human environment without active domestication." Rats invade when ecosystems get disrupted. In terms of the bare necessities, "rats need only a place to build a burrow (usually open soil but sometimes within buildings or piles of material), access to fresh drinking water, and around 50 grams of moderately calorie-rich food each day," according to Matthew Combs, a doctoral student at Fordham University who is studying the genetic history of rats in New York City. In a human-dominated landscape like New York or Vancouver, "It comes down to where rats have found a way to access resources, which often depends on how humans maintain their own environment."

It's not hard to understand why humans often think of the rat lifestyle as a parasitic response to our own. But that's not entirely true. "I have to stop myself sometimes because I want to say that rats have adapted to our cities," says Combs. The reality is that rats were perfectly positioned to take advantage of the disruptions caused by human settlement long before we arrived. They've been on Earth for millions of years, arriving long before modern humans evolved, about 200,000 years ago. Before cities were even a glimmer in our eye, rats were learning to become the ultimate opportunists. "They were likely stealing some other species' food before ours," Combs says. Even in the still-remote mountain habitats of New Guinea, says Aplin, "you tend to find rats living in landslides or along creek systems where natural disturbance is going on." Walk into a lush, primary, intact forest, "and they're pretty rare." It's not that rats have become parasitic to human cities; it's more correct to say they have become parasitic to the disturbance, waste, construction, and destruction that we humans have long produced.

Which brings into question the constant human quest to disrupt rats and their habitats. As much as rats thrive in disrupted environments, Byers says, they've managed to create very stable colonies within them. Rats live in tight-knit family groups that are confined to single city blocks and that rarely interact. The Rat Project hypothesized that when a rat is ousted from its family by pest control, its family might flee its single-block territory, spreading diseases that are usually effectively quarantined to that family. In other words, the current pest control approach of killing one rat per concerned homeowner call could be backfiring, and spreading disease rather than preventing it.

The diseases that rats might be spreading aren't just their own. Himsworth likes to say that Vancouver's rats are like sponges. Their garbage-based diets allow them to absorb a diverse collection of bacteria that live throughout their city, in human waste and in our homes. "So it's not like the presence of harmful bacteria is characteristic of the rats themselves," she says. They get that bacteria from their environment, and when they move, they take these place-specific pathogens with them.

When "stranger" rats come into contact, Byers says, territorial battles ensue. "They urinate out of fear and they draw blood," she says—perfect for expelling and acquiring even more bacteria. It's during these territorial brawls, Byers and her colleagues believe, that bacteria can converge, mix, and create new diseases. "The rat gut acts as a mixing bowl," says Himsworth, where bacteria that would otherwise never interact can swap genes and form new types of pathogens.

One example is a strain of methicillin-resistant *Staphylococcus aureus*, or MRSA, that Himsworth found in V6A's rats. It included a piece of genetic material from a very closely related superbug called methicillin-resistant *Staphylococcus pseudintermedius*, or MRSP, which is often only associated with domestic animals like pet dogs. It seems that rats pick up human MRSA from the sewers or the streets and canine MRSP from our yards, then mix them in their guts. These new human-rat bugs could then potentially spread back to people via the rats' droppings and saliva.

*

In V6A it's hard not to notice the litter around us. Garbage has bubbled out from under the lids of trashcans, and a pile of empty syringes surrounds a parking-lot trap. Walking across this landscape of debris, cracked concrete, and weeds, Byers stops at another trap, which is set on what she has named "Block J." She and two student assistants are heading the project's second phase, which involves tracking the real-time movement of rats, using ear tags. Once these trees are mapped, she will begin to euthanize individual rats and see how their family responds. Part of her PhD work is to understand how human-caused disruptions, pest control in particular, affect how rats move throughout V6A. The hypothesis is that the disruption will send communities scurrying for new ground. With nearly 100 cages to check today, Byers moves hastily to a trap on Block 8. No rat here, but this one did catch a skunk.

A significant finding from the project's original phase, Byers tells me, is that not every rat in V6A carried the same disease. Rat families are generally confined to a single city block, and while one block might be wholly infected with a given bacteria, adjacent blocks were often completely disease-free. "Disease risk doesn't really relate to the number of rats you're exposed to as much as it does to which family you interact with," says Robbin Lindsay, a researcher at Canada's National Microbiology Lab who assisted the Vancouver Rat Project screen for disease. If those family units are scattered, diseases could potentially spread and multiply—something Byers is hoping to figure out through her PhD work.

If that's true, a city's rat policy should include doing the unthinkable: intentionally leave them where they are. "It might be better to maintain local rat populations that already have some sort of equilibrium with the people who live there," says Aplin. Many of the diseases that we share with rats are already part of a human disease cycle established over centuries, he says. Seen this way, rats are irrepressible—"a force of nature, a fact of our lives." Rather than focusing on killing them, we need to try to keep their populations stable and in place—and that includes managing rat immigrants.

An established rat society in a neighborhood makes it a much less viable destination for other rats, for example those entering through ports. Exotic rats can be more of a threat than those adapted to the region because each rat community evolves with its own suite of unique pathogens, which it shares with the other vertebrates in its ecosystem. New rats mean new diseases. The big question now, Aplin says, is "what happens when these different pathogens come together? This is something that I'm just starting to think about now. If the local rat population is suppressed, if you're actively getting rid of it, then you're also actively opening up niches for these foreign rats to enter."

In Vancouver, this is a fact of life. "One important thing we do have right over there," says Byers, motioning with her left hand, "is Canada's largest shipping port." Vancouver sits on Vancouver Harbor, which houses the great Port of Vancouver. In one of Himsworth's earlier studies, she found mites on the ears of rats that live by the port and compared them to rats that take up residence around V6A. Port rats had malformed ears full of a strange breed of mite previously unknown to Canada—"an exotic species that's found in Asia," Himsworth says, which happens to be where Vancouver gets the majority of its imports. These foreign ear mites were not found on rats from any other block.

"So I think Aplin's theory has a lot of merit," Himsworth says. "It seems that the established rat population at the port acts as a buffer." Himsworth wonders if this is precisely what has kept an otherwise highly contagious mite from spreading throughout V6A.

Disruption, of course, doesn't come from just ports and pest control. It is part and parcel of modern civilization. Vancouver's population is growing steadily (by about 30,000 residents each year), bringing housing development, demolition, and more garbage. Even our love of birds can be a problem. Two years ago, for example, rats invaded a playground and community garden in East Vancouver, a bit outside of V6A. Several media sites reported on the visitors, which were evidently drawn in by birdseed dropped by a single individual. The area soon became known as "Rat Park." The City of Vancouver urged the garden's coordinator to put up signage asking people to avoid feeding the birds and to pick up their overripe vegetables. An exterminator was hired as well—adding more disruption still.

Himsworth hopes the new science will sway Vancouver's existing policy on rats, which, she stresses, is currently "essentially nonexistent." This bothers her a lot. "I know that Vancouver Coastal Health essentially has the standpoint that, 'Well, we don't see the disease in people so we don't worry about it,'" she says of the region's publicly funded health-care authority. Homeowners with

rat infestations can ring 311, Canada's 411, to report an infesta-
tion, but that's not a preventative response. "Rats are pests, and we
don't spend health-care dollars to track pests," said media officer
Anna Marie D'Angelo of Vancouver Coastal Health. It was a message echoed by issues management communications coordinator
Jag Sandhu of the City of Vancouver: "The City of Vancouver does
not track the rat population." To Himsworth, this is shortsighted.
"They're not taking the rat disease risk seriously because they
haven't seen it in humans yet—but that's not where diseases start."
She also believes the issue is in part one of social justice. Rats typically affect poor areas, like V6A, that have little political clout.

Back inside one of Byers's traps in V6A, needlelike nails are
lightly scraping on the metal. "It's a black rat," Byers tells me—the
famed carrier of the Black Death. Byers says she isn't concerned
about bubonic plague, which in North America is mainly carried
by prairie dogs. But there were 13 rat-driven bubonic plague out-
breaks in seven countries between 2009 and 2013. And there are
plenty of new diseases cooking.

ROBERT DRAPER

The Battle for Virunga

FROM *National Geographic*

WHEN THE RANGER studied the ragtag crew he was supervising, seven young men repairing a rugged road that leads to Virunga National Park, it did not take much to see what he had in common with them. They were all born and raised in or around the park on the eastern edge of the Democratic Republic of the Congo. None of them were rich. None of them would ever be rich. All of them had seen loved ones fall by the capricious machete stroke of a war with murky logic and no foreseeable end.

And now here they all were, working for the park, filling potholes and clearing drainage ditches in the furtherance of something considerably more profound than nine miles of rough gravel. The road joins the Bukima ranger post with tourists from the West, whose money helps support Africa's oldest national park. These visitors come here principally to fulfill a dream—namely, to stand mere feet away from the park's illustrious residents, the rare mountain gorillas.

Less famous but just as important, the Bukima road connects farmers outside the park with village markets and the city of Goma beyond. For years it had been a morass of large rocks and quicksandlike mud. Its impassability made hard lives that much harder. But now the park was pouring money into the road's reconstruction. And local men like these were repairing it. So the road also constituted a bond, albeit a slender one, between the region's most visible national institution and villagers who view the park with hostility, and at times rage, believing the land should still belong to them.

Here was where the ranger, a captain named Theo Kambale, parted ways with the young men. Kambale's heart held nothing but reverence for the park. You could see it in the crispness of his uniform, the care with which he tucked his green pants into his boots, which he fastidiously polished. Kambale was 55 and had spent 31 of those years as a ranger. His father, also a ranger, had died in 1960, the year of Kambale's birth, gored by an African buffalo. His older brother had also been a ranger. He too had been slain in the line of duty, in 2006. The killer was not a wild animal but instead a member of one of many armed groups that have ravaged and occupied Virunga for two decades.

To these young men raised in poverty, that Virunga's tremendously fertile soil, its trees, and its creatures should be protected by law for the viewing pleasure of well-off tourists struck them as a grave injustice. They were swept into a militia known as M23, which touted a host of grievances against the corrupt government but in the meantime was content to loot and rape its way through a slice of eastern Congo near the park's southern sector. By the end of 2013, after more than a year and a half of fighting, the Congolese Army, backed by United Nations troops, routed M23. Among the militia's foot soldiers deemed salvageable, by UN peacekeepers and park officials, were these seven.

The work on the Bukima road was harder and less profitable than looting. But the former rebels kept at it. Kambale was impressed. He talked to them from time to time. "Before now, all you were creating was insecurity in the region," he would say. "Now you're building this road. It's a start. From here you can go on to do other things. But you can't progress if there's no security. So tell that to your friends. Tell them to leave their armed groups. Because that is not life. This"—and he would gesture toward the road—"is the beginning of life."

The ranger hoped that his message would sink in. He knew of their desperate backgrounds. He was aware that most had been conscripted by force. Across their arms and backs was a grisly network of scars, testifying to their semi-enslavement. Seeing these men in their 20s permanently marked by brutality, Kambale thought of his own injury, delivered by a militia spear to his right leg. Proof of residency, you could say. If they could look past their battle wounds, perhaps this park could be saved.

*

There is no nationally protected area in the world quite like Virunga, in ways both blessed and cursed. Its approximately 2 million acres include a web of glacier-fed rivers, one of Africa's Great Lakes, sun-bleached savannas, impenetrable lowland rainforests, one of the highest peaks on the continent, and two of its most active volcanoes. Virunga hosts more than 700 bird species (among them the handsome francolin and Grauer's swamp warbler) as well as more than 200 mammals (including the odd-looking okapi, with zebra-striped hind legs, and 480 of the world's 880 remaining mountain gorillas). Standing where the Semliki River flows out of Lake Edward with the Rwenzori Mountains glowering in the distance, serenaded by a moaning Greek chorus of water-besotted hippos, and gazing down at a thoroughly uncontaminated tableau of swimming elephants and strutting saddle-billed storks backlit by a low morning sun, one becomes very small, very silent, and very aware that nature's brave feint of indomitability has all but come to an end.

For Virunga has been, going on two decades, a war zone. In 1994 the horrific ethnic conflict in neighboring Rwanda that led to the genocide of Tutsis by Hutus spilled across the border into Congo. Hutu fighters and more than a million refugees fled Rwanda after their defeat, settling in nightmarishly overcrowded camps around the park. Some Hutus later formed the Democratic Forces for the Liberation of Rwanda—known by its French acronym, FDLR—the militia that killed Kambale's older brother. Congolese Tutsis eventually responded with the National Congress for the People's Defense, or CNDP, which then spawned the March 23 movement, or M23. One bloody iteration after the next—fomented by these armed groups—has plowed into the park like a threshing machine.

Many of the fighters, along with Congolese Army soldiers purporting to defend the territory, lingered well after the cease-fires, expunging the park's wildlife for personal consumption or for sale as bush meat. Thousands remain in the jungle to this day, and thousands more from a shifting array of locally formed militias called Mai-Mai have joined them. Attempts by rangers to drive them out have led to deadly reprisals. This past March two rangers were executed in Virunga's central sector, driving up the death toll of park rangers to 152 since 1996.

A different kind of war also looms over Virunga. This one pits the park and its ecological well-being against the search for oil. London-based Soco International obtained a concession in 2010 that allowed it to explore about half of Virunga, including the area near Lake Edward. After a sustained outcry led by conservation groups, four years later Soco backed down and now says it no longer holds the concession. The Ugandan government, however, has shown an interest in exploring for oil on its side of the lake, a grim reminder that the park and its precious resources are anything but sacrosanct.

The park is also a volatile staging ground for Congo's internal grievances. As it happens, Virunga's terrain is among the most fecund in Africa. That it has been set aside for conservation since the park's founding in 1925, thereby depriving one of the world's most deeply impoverished populations of badly needed natural resources, stokes seething discontent among the area's 4 million inhabitants. Many, in defiance or ignorance of the law, cut down the park's trees for charcoal, plant crops in its forests, kill its wildlife. Some form Mai-Mai militias and take over sections of the bush, emerging in periodic sprees of violence. Others run for elective office essentially as park abolitionists, vowing to reverse the misdeeds of the Belgian colonizers who they say tricked the locals into selling their treasured farmland, or so goes their campaign narrative.

This pervasive climate of resentment is not a small misfortune. Rather, it represents an existential challenge for Virunga. "The truth is that we're not going to succeed unless we mobilize a critical mass of funding," the park's director, Emmanuel de Merode, said, noting that the land, if it were developed, would bring the communities about $1 billion a year. "Unless we equal that, this park won't survive."

Owing to the region's chronic instability, a mere one-tenth of Virunga is accessible to visitors—and really only half of that could be described as tourist-friendly. The park's VIPs—the 250 to 300 mountain gorillas that are habituated to humans—are kept under daily watch by a security team of 80 humans, as would befit a president or a pope. Virunga is national property, but the government in Kinshasa contributes only 5 percent of the park's $8 million annual operating budget. Most comes from the European Union, the U.S. government, and international nonprofits. Though a first-class hotel, the Mikeno Lodge, opened in 2012 near the gorilla sector, and the sumptuous tent camp on Tchegera Island in Lake

Kivu began receiving guests in 2015, the number of visitors has not come close to matching that of the park's prewar heyday. Indeed, the lodge was empty throughout much of 2012 and 2013 as Virunga hosted the latest season of bloodshed, the M23 rebellion.

In the years since, the park has experienced a renaissance thanks to projects such as the Bukima road-building effort, which aim to show Virunga's neighbors that respect for the park will be rewarded. In particular, de Merode, with substantial support from the Howard G. Buffett Foundation, has embarked on an ambitious $166 million hydroelectric scheme utilizing the park's rivers, with the aim of electrifying one-fourth of the area's households by 2020 and creating 60,000 to 100,000 jobs along the way. The outcome, de Merode hopes, will be peace—and with that, more tourism, and thus more income for the region's people, spurring an altogether different cycle from the one that is still bedeviling eastern Congo.

Meanwhile, slowly, the wildlife has begun to rebound. Since the massacre of seven mountain gorillas by charcoal traffickers in 2007, their population has been rising. In the central-sector preserve known as Lulimbi, hippos have mounted a surprising recovery, while elephants are wading back across the Ishasha River from the safe haven of Uganda. Aggressive antipoaching operations by rangers have sent an unambiguous message to ivory and bush-meat traffickers: Virunga is no longer an anything-goes playground.

"It was a beautiful place," Kambale said one afternoon as he stepped carefully through the weed-choked ruins of the Rwindi Hotel in the central sector. "The hotel was always over capacity. Everyone came to see the wildlife and take pictures. There were so many animals. Even the parking lot was full of antelopes and wild pigs and all types of monkeys."

Today only baboons clamber through the brush. The cylindrical bungalows, the restaurant, the ballroom, the pool where *mzungu* ladies sunned themselves on hot days like this—all vacant and caked with two decades' worth of neglect. The ranger wore a doleful smile, and his eyes were lost in the past. He was born and raised near the Rwindi patrol station. During the year of Kambale's birth, 1960, Congo won its independence from Belgium. Its population, 15 million, was a fifth of what it is now. There was plenty of land to go around, for farmer and animal alike. As a young ranger in the 1980s, Kambale sometimes had to climb a tree to avoid being trampled by a buffalo. When the dictator Mobutu Sese Seko came to visit—to entertain guests, to plot a course for the country he had renamed Zaire, but most of all to fish on the Rwindi River—it was Kambale's job to hook a live worm onto Mobutu's line. "Mobutu had great respect for the park," said Matthieu Cingoro, a lawyer for the Congolese national park system. "No one could farm in it or cut down trees. No one would even dare trespass."

Then came the refugees from Rwanda. The Rwindi Hotel abruptly locked its doors. The patrol station now saw a desperate new breed of visitor. "There were many of them, and some had guns and ammunition," Kambale remembered. "Like that, the population increased, and these people had no food and had to look for charcoal, wood for fire, even meat in the park." One armed group begat another. The distinctions blurred. Congolese soldiers deserted their posts and disappeared into the bush. Some joined Mai-Mai militias, which at times confederated with the Hutu-based FDLR against all comers, including the rangers who sought to deny them a livelihood inside the park.

As the Mobutu regime collapsed in 1997, so did any semblance of governmental structure. Virunga's rangers saw their salaries slashed. They had to fend for themselves. Many did so by taking money from poachers, who would brazenly call a compromised ranger and direct him to come pick up a slaughtered buffalo. Other rangers distributed tickets to locals, allowing them to harvest wood for charcoal with the understanding that a generous slice of the profits would be handed over to Virunga's uniformed men—and make its way up the food chain.

Even in this moment of relative calm, ghosts have claimed far more of the central sector than its decaying hotel. The former ground zero for park tourists, Rwindi station, is still a no-go zone. The walls of the sector commander's office are pocked with bullet holes. A UN military base lies nearby. Signs posted throughout Rwindi urge the locals to report any signs of a ranger's suspicious activity.

Late one morning Kambale and two other armed rangers drove me to Vitshumbi, a village on the south bank of Lake Edward, inside the park's boundaries. Conceptually Vitshumbi is a fishery with 400 boats licensed to fish on the lake, supporting about 5,000 people. In reality Vitshumbi is a squalid town with thousands of boats and perhaps 40,000 residents with no electricity or running water.

What it does have are Mai-Mai militias, which have offered protection to Vitshumbi's fishermen and farmers in exchange for a surcharge. Behind the militias, Kambale and other rangers say, are politicians who supply the outlaws with boats and weapons. "It used to be that the Mai-Mais just fought with spears and machetes," a young ranger stationed in Vitshumbi told me. "Now the politicians have given them guns." The ranger pointed to a bullet scar on his left biceps, a souvenir from a recent encounter with Mai-Mais on Lake Edward. One ranger and seven Congolese soldiers had been killed.

Elsewhere during my three weeks in the park, unrest flickered ominously like a rogue torch in the night. From Vitshumbi a ranger boat was waiting to take me north to the hippo enclave of Lulimbi. Minutes before embarking I learned that my trip was canceled by the park's director of security, who called to say the lake was not considered safe from attack. Three days before that, in the southern sector where the mountain gorillas reside, an angry phalanx of at least 300 villagers had blocked the road outside the Mikeno Lodge for hours, saying that the park had failed to compensate them for cutting down some of their trees that would have interfered with newly installed electrical lines. Adding to the villagers' disquiet was the fact that a thousand or more Rwandan Army soldiers had quietly crossed the border to hunt down FDLR fighters. A week later, upon arriving in the northern sector, I watched as a squadron of rangers and Congolese soldiers made out for Mayangose, northeast of the city of Beni, where they forced out an encampment of 800 squatters who had been egged on by politicians to seize parkland.

A few hours after Kambale had escorted me from Rwindi to Vitshumbi in a park jeep, the central sector's accountant left Rwindi for the day and drove home on his motorbike along the very same road—only to be waylaid by three men who jumped into his path and pointed Kalashnikovs at his chest. They tied his hands and dragged him off into the bush. Later that evening the accountant's family received a call demanding a $5,000 ransom.

Word reached park headquarters. More than a hundred rangers and Congolese soldiers were dispatched to the central sector, along with aerial reconnaissance and tracking bloodhounds and spaniels. The dogs located the accountant's scent. The pursuers set up a perimeter and began firing shots into the air. The kidnappers fled. Ambling through the bush, the accountant came upon the welcome sight of his fellow park employees. It was, for him, a harrowing ordeal—but also a show of swift action by de Merode, the man Kambale refers to as "our only hope."

As heroic figures go, the 46-year-old Emmanuel de Merode seems somewhat miscast. Milky-faced, thin, and mild in manner, the Virunga director and chief warden does not exactly fill a room; he does not even fill his uniform. When I first met him, at a National Geographic Society event in Washington, D.C., I was sure he must be someone else, waiting alongside me for the actual de Merode to materialize. By ancestry he's a Belgian prince, a title bestowed on his family because a forefather helped the country gain independence from the Netherlands. From this limited appraisal, one could imagine de Merode best suited to a life beside a fireplace with a glass of burgundy, sweater-clad, rather than in one of the world's most notorious conflict zones. But de Merode was born in Africa, spent his youth in Kenya, trained as an anthropologist, and has worked his adult years in conservation, much of it in Congo.

Beneath de Merode's baggy ranger shirt are two sets of entry and exit wounds; one bullet went through his left lung and the other his stomach. He acquired these injuries in April 2014, while driving from Goma back to the park on a deserted and poorly paved stretch of marshy road about three miles south of Rugari. The would-be assassins were never found. (The investigation, his associates note with fatalistic eye rolls, is ongoing.) News of the shooting descended upon eastern Congo "like a thunderclap," recalled Kambale. Today de Merode's friends notice the occasional cough—the only utterance of lingering discomfort.

De Merode became Virunga's director in 2008, at the park's precise nadir. The previous director had been arrested earlier that year and accused of participating in a charcoal-trafficking ring and planning the gorilla massacre. (He was not convicted, for lack of evidence.) About six months earlier, the park's new occupier had become the CNDP, a militia backed by Rwanda to take on the FDLR. De Merode's first order of business was an audacious act— to show up unarmed at CNDP headquarters to ask that his rangers be permitted to return to the park. The militia's leader, Laurent

Nkunda, granted the request. De Merode then set to work cleaning up the ranger force. He slashed its ranks from 1,000 to 230 (later bringing the number back up to 480, including 14 women) and hiked monthly salaries from a pitiable $5 to a decent living wage of $200—"enough," he said, "to justify a zero tolerance of corruption."

De Merode then began trying to improve relations with the local population. He listened to the people's complaints. For decades the park had promised that half of each tourist dollar would go back to the community. Where was that money being spent? The roads, the schools, the hospitals were steadily deteriorating. Meanwhile elephants were destroying crops.

"Before de Merode started showing up, we didn't even know the park had a director," a fisherman in Vitshumbi told me. "Now you see the rangers have clean uniforms, good weapons. You see what a difference he's making." The director even sat down with Congolese militia groups—though with mixed results. "If we can have a constructive dialogue with militias that keeps people safe and keeps rangers from being killed, we're willing to do that," de Merode said. "But often it's been disappointing, because it hasn't been an honest dialogue."

Regardless, his presence has registered with his adversaries. In 2012 a ranger major named Shadrack Bihamba was cornered by Mai-Mais on the shore of Lake Edward and led at gunpoint into the bush. Bihamba said the militia's leader was worried, telling the others, "He's an officer. If we kill him, de Merode will move heaven and earth to annihilate us." He instructed his men to release Bihamba. "Even though they're Mai-Mais and have their strength in the bush," Bihamba said, "they still fear de Merode, because they know he has the entire population behind him."

Still, de Merode knew something that some of his enemies did not—which was that his growing prominence alone could not sustain the park. It needed money to enforce the law and make permanent allies out of the park's neighbors. The only way to achieve the latter, de Merode concluded, would be "to use the park as a basis for creating mass employment, but in a way that wouldn't damage the park." That goal led him to the park's northern sector —and specifically to the Butahu River, which cascades from the glacial peaks of the Rwenzori Mountains into the outskirts of Mutwanga village, a typically meager community that lacked electricity.

In 2010 the park began hiring villagers to dig canals and lay the foundation for what would become Virunga's first hydroelectric plant. For $110 the park would connect a Mutwanga household, which could then buy electricity on a modest pay-as-you-go basis. In 2013 the power came on, and de Merode held his breath.

I had not seen Mutwanga before it had electricity, and it hardly resembled a boomtown when I spent a day touring the mud-splattered village. Still, the residents spoke of the change as transformative. What it had cost in a single day to power their shops in generator fuel now bought an entire month's worth of electricity. Students could do their homework in the evening. The hospital functioned at all hours. People were buying irons, televisions, and CD players. The owner of a computer-repair store was renting out DVDs and preparing to open the town's first Internet café, so that villagers would no longer have to drive an hour to Beni to send an email. A couple from Beni actually moved to Mutwanga in 2014 to realize their dream of owning a small printing shop. All of this despite the fact that only 500 of the community's 2,500 households have been hooked to the hydroelectric plant's modest 400-kilowatt output. And while de Merode's team makes plans to accommodate the long waiting list, in April a factory powered by the park began making soap. It employs about a hundred workers from the area. "Mutwanga became our laboratory test," de Merode said.

A second, larger hydroelectric plant came on line in December, and by the end of 2018 two others should be running. Those four plants would bring de Merode halfway toward his goal of producing a hundred megawatts of power. Selling that electricity, he predicts, would "enable us to ensure that the park will be financially sustainable for the next one hundred years." Enough additional revenue would be generated to invest millions a year in community projects and conservation efforts in other Congo parks.

De Merode's expectation is that electricity will catalyze economic development. "The reason there isn't industry is there's no access to cheap energy. That's really what the park can offer," he said. That this will lead to a flowering of entrepreneurship is far from a sure thing. "There aren't any business role models here," the soap factory's managing director, 29-year-old Leonard Maliona, told me. "Young people have nothing to aim for, other than being a politician or joining a militia."

The notion of Virunga as the region's "economic engine," to

use de Merode's terminology, conjures up a spectacle that some may find unusual. Among other things, the scenario suggests that Congo's leaders have essentially consigned the fate of one region of their country to a single park and its director, who shares his Belgian heritage with the country's former colonial power. It also risks replacing a population's lingering hostility toward the park with an intense dependency on it. De Merode's gamble is unapologetically high stakes. And it rests heavily on young men who agree to beat their swords into plowshares and do an honest day's work on little farm roads like the one up to Bukima.

The two laborers, both in their mid-20s, wore fluorescent orange vests over their T-shirts and were filling potholes on a shady stretch of the road. The taller of the two, with hooded eyes, was named Bushe Shukuru; the shorter one, with a quiet but easy smile, went by Gato Heritier. The two were childhood friends. Each time armed insurgents came, causing the villagers to run for miles until the sounds of gunfire had diminished, the teenagers would make it a point to search for each other at the refugee camps. On separate days during the spring of 2013, first Shukuru and then Heritier were caught in their village by M23 soldiers who tied their arms and marched them off to the place where they again found each other: the military base in Rumangabo, near the park's southern sector, which had been taken over by M23. They joined a thousand or so young men who were also involuntary conscripts at the rebel faction's training camp.

The commanders told Shukuru, Heritier, and the others that the government had failed eastern Congo. With proper training, they said, M23's new warriors would take over the region and then advance westward and conquer Kinshasa. They were taught how to shoot, march in formation, attack, and withdraw. For their shortcomings, they were beaten with wooden sticks—some of them to death, right in front of the others. Others died of starvation from the paltry daily rations: a single cup of cornmeal. Three months of this, and then Shukuru and Heritier were sent to battle. By November both could see that M23 stood no chance against the army and UN forces. They fled and found each other again that month, in a UN compound.

Now here they were, in matching vests. The road they tended was, by rural Congolese standards, almost sleek. Yam and corn farmers, cattle and goat herders, schoolchildren and churchgoers negotiated the sloping path in half the time it once took. "The road's really had a big impact," said Heritier as he sat on a log and wiped the sweat off his face. Shukuru agreed: "That's why I don't mind doing this job. You can tell it's helping this community."

But, they acknowledged, $3 a day for eight hours of backbreaking work was not where they saw themselves for long. As a small child, before he understood what life in eastern Congo had to offer, Heritier had imagined himself as "some kind of big guy. A doctor. Maybe even the president. I mean, why not?" If he saved his money, perhaps he could be a mechanic, and Shukuru might one day open a shop of some sort. A small and quiet but honest and peaceful destiny that began with this road, leading uphill to the mountain gorillas. From there the peace would spread northward to Rwindi station, where Theo Kambale was also daring to harbor modest dreams. Recently, he had heard, a lioness and her cub had been spotted watering themselves on the banks of the Rutshuru River. And he had heard something else—that along with the slow return of wildlife, the long-abandoned Rwindi Hotel may also return, if the park can find the money to restore it.

It was, as Kambale told the young men on the road, the beginning of life.

TOM KIZZIA

The New Harpoon

FROM *The New Yorker*

THE SPRING HUNT started promisingly last year for the village of Point Hope, on the Chukchi Sea in northern Alaska: crews harpooned two bowhead whales and pulled them onto the ice for butchering. But then the winds shifted. Out on the pack where the water opened up, the ice at the edge was what is called *sikuliaq*, too young and unreliable to bear a 30-ton whale carcass. The hunters could do nothing but watch the shining black backs of bowheads, breathing calmly, almost close enough to touch.

On a trip to the ice edge, Tariek Oviuk, a hunter from Point Hope, felt a strange sensation: the lift of ocean waves beneath his feet. The older men, nervous about the rising wind, hurried back toward shore, but the younger hunters remained, stripping blubber from a few small beluga whales. Then the crack of three warning shots came rolling across the ice, and the hunters scrambled for their snowmobiles. "As soon as we heard those shots, my heart started pounding," Oviuk recalled.

As Oviuk told me the story a few months later, we were sitting in the kitchen of his friend Steve Oomittuk, a former village mayor, eating strips of *maktaaq*—chewy beluga blubber—off a piece of cardboard that quickly grew sodden with whale oil. Oviuk is 35, tall, and square-jawed, a former basketball star for the Point Hope high school team, the Harpooners, and a member of a local troupe that performs traditional storytelling dances. "That was our way of communication," he told me. "That was our people's iPhones since time immemorial."

Oviuk said that when he heard the shots he started running,

then jumped into a passing sled filled with slippery blubber. "That's not a beautiful thing, to be in a sled full of *maktaaq*," he said. Another snowmobile driver swung by to rescue him, and Oviuk scrambled aboard. Then they stopped: a gap of blue water 100 feet across had opened between them and the shore-fast ice. The driver, in a parka and ski pants, said, "Hold on." Accelerating, their heavily laden snowmobile leaped off the ice and skipped over the surface of the Chukchi Sea. Others followed, engines screaming, until everyone was across. "I didn't believe in global warming —I'll tell you that straight up," Oviuk said. "But I teared up out there. I was thinking, 'Every year we don't know if it's the last time we're going to see the ice.'"

Point Hope sits at the northwesternmost corner of North America, on one of the oldest continuously settled sites on the continent. Eight hundred people live near the eroding tip of a 15-mile gravel spit thrust into the Chukchi Sea, a peninsula that the Inupiat call Tikigaq, or "index finger." For 2,000 years the digit, stuck into coastal migration routes, has provided an ideal hunting perch. Tikigaq was a capital of the precontact Arctic, whose prosperity depended on a subtle understanding of the restless plains of ice that surrounded the community in winter.

In Paris last winter, 195 nations agreed to limit greenhouse-gas emissions and slow the warming of the planet. President Obama, speaking at the Paris conference, called for the global economy to move toward a low-carbon future, citing his own recent trip to Alaska, where melting glaciers, eroding villages, and thawing permafrost were "a glimpse of our children's fate if the climate keeps changing faster than our efforts to address it." The goal in Paris was to hold the average global increase in temperature to less than two degrees Celsius. The Arctic, which is warming at twice the rate of lower latitudes, has already shot beyond that: average annual air temperatures have increased by about three degrees. If trends continue, northern Alaska is expected to warm another six degrees by the end of the century.

These days the ice disappears so fast in spring that villagers struggle to catch bearded seals, whose skins are traditionally used in Point Hope to cover hunting boats. Ice cellars in the permafrost, packed with frozen whale meat, are filling with water. People are worried about these changes. Like most families in the village, Oomittuk's survives on wild game; much of the living space in his small house was taken up by two big chest freezers. Villages in Alaska's Arctic consume nearly 450 pounds of wild game and fish per person each year, according to a recent study. "Without the animals, we wouldn't be who we are," Oomittuk said.

With a warm July wind battering the peninsula, Oomittuk took me on a four-wheeler ride for a glimpse of Point Hope's past glory. Years ago elders on the tribal council had picked Oomittuk as a kind of tradition-bearer. An amiable 54-year-old with a long wisp of chin hair, he had grown rounder and softer since his own whale-hunting days, when he once helped repel a polar bear nosing into his tent by brandishing a cast-iron skillet. He explained that the lumps in the tundra, visible in all directions, were the husks of prehistoric earthen homes.

Along the coast where people hunt and camp, Oomittuk said, there are haunted places where no one ever stops. (In 1981 the ethnographer Ernest Burch identified four such zones, avoided because of "nonempirical phenomena.") Explorers and whalers in the 19th century described Point Hope as an open graveyard, with skeletal remains arrayed for miles atop funerary racks of bleached whalebones—essential building materials in a land without trees. Episcopal missionaries at Point Hope eventually persuaded villagers to bury the human remains—as many as 1,200 skulls, according to one account—in a single mass grave, surrounded by a picket fence of repurposed bowhead mandibles. At an abandoned village site nearby we found a line of the weathered-gray bones, staked into the tundra by missionaries a century ago to help converts find their way through the blizzards to church.

We drove to the beach overlooking the Chukchi Sea, where the evidence of erosion was plain. The peninsula used to extend considerably farther out. Prehistoric settlements have eroded away, and artifacts wash up after fall storms. "I love my way of life," Oomittuk said in a soothing baritone. "My grandfather's life. The cycle of life. The connection to the land, the sea, the sky."

Few Americans are as bound to the natural world as the whale hunters of the Arctic, or as keenly affected by the warming atmosphere. Yet few Americans are so immediately dependent on the continued expansion of the fossil-fuel economy that science says is causing the change. The underground igloo where Oomittuk was born, in 1962, had earthen walls braced with wood scraps and whalebone, and a single electric light bulb. Point Hope today is a grid of small but comfortable homes laid out around a new school and a diesel-fired power plant—everything provided by a regional municipality with 8,000 permanent residents and an annual budget of $400 million. Oil drilling in the Arctic has paid for nearly all of it, and Oomittuk does not want to go back.

There is a cost, though. Over the horizon from the beach where we stood that summer, Shell Oil had assembled a floating city. The project was opening an entirely new part of the Arctic Ocean to oil drilling. The dangers posed to the Tikigaq hunting culture by a massive spill were never far from Oomittuk's mind. But he worried too about how the village would survive if there were no more oil industry. The tradeoffs have racked Alaska's Inupiat communities. For nearly a decade Point Hope pressed a lawsuit against the offshore leases, becoming the last stronghold of indigenous opposition. Finally, in the spring of 2015, the village dropped the suit. On the day the thin ice nearly carried Tariek Oviuk out to sea, his whaling captain had been in Houston, meeting with Shell officials.

The first oil boom in the Alaskan Arctic was devastating for the Inupiat. It began in 1848, when Yankee whalers, having depleted the sperm whales of the Pacific, discovered an unexploited population of bowheads north of the Bering Strait. In two decades the fleet killed nearly 13,000 of the oil-rich whales, and then it turned to decimating the walrus. Eskimo hunting communities, already struggling with alcohol and diseases brought by the whalers, faced another scourge: hunger. In the early 1880s a government revenue cutter that landed on St. Lawrence Island, south of the Bering Strait, found that a thousand inhabitants had died of starvation. At Point Hope dozens of people starved, but only after eating their dogs and making soup from the skins off their boats.

The second boom came after 1968, when oil was discovered at Prudhoe Bay, and this time the Inupiat were better prepared. A pipeline had to be built across Alaska to a tanker port, and the Inupiat, along with other Alaska natives, asserted rights to the federal land along the way; the question of aboriginal land rights had gone unresolved since Alaska was bought from Russia.

In 1971 Congress awarded the natives a huge settlement: 44 million acres and nearly $1 billion. In another age the settlement might have been used to create reservations, to sequester aspects of a traditional life. Instead the land went to 12 new for-profit corporations owned by native shareholders. Some natives were ambivalent about entrusting their future to a corporation and worried about losing hunting and fishing grounds to sale or bankruptcy. But the more urbanized leaders saw a means of forcing their way into Alaska's modern economy: one activist said that native corporations would be "the new harpoon."

When Steve Oomittuk was growing up, Point Hope, the once-great capital of the Arctic, had receded to the margin of the civilized world. People lived in small frame houses and in a few last underground homes, scavenging materials by dog team from abandoned whaling and military sites. Oomittuk recalled that the most exquisite treat available at the village store was a roll of Life Savers.

In the decade after oil was discovered, regional leaders organized a municipal government and set out to reverse a long history of neglect. The North Slope Borough, encompassing the town of Barrow and seven small villages in an area the size of Minnesota, was given authority to collect property taxes from the new production facilities and pipelines. The borough built power plants, schools with swimming pools, sewer systems with heaters to prevent freezing. Today the government subsidizes a tribal college, child care, bus service, heating oil, and a $35 million public safety department. In 1997 the borough's helicopters rescued 173 whalers drifting into the fog on a breakaway slab of ice.

In Point Hope, Oomittuk's father served on the local tribal council and helped launch the new government. The entire community was moved two miles from the fast-eroding tip of the peninsula, and the population doubled as wages and transportation lifted the air of deprivation around village life. Oomittuk began working as a carpenter and started a family.

He was uneasy about some of the changes that the new prosperity brought. During a no-bid construction boom in the 1980s, he watched a corruption scandal bring down a borough mayor. New tools like outboards and snowmobiles improved hunters' productivity but required cash; Oomittuk had one of the village's last dog teams, until a power line blew down and landed in his dog yard. On the other hand, Oomittuk was on the borough's payroll for a decade as village fire chief—one of many positions that set North Slope communities apart from the 200 or so other villages in Alaska. Point Hope today has a spotless fire station, with a full-time staff of four and a fire engine, a tanker truck, and an ambulance. By contrast, when a fire last spring in Emmonak, a village in the southwestern part of the state, roared through a fish-processing plant, residents could only stand beside their broken-down equipment and watch.

The cultures of the Arctic were known for being quick to adopt new technology, but subsistence-hunting traditions remained at the heart of Inupiat life. Oomittuk joined the tribal council and worked as a harpooner in his uncle's *umiak,* or skin boat. In the North Slope villages, whaling captains continued to serve as leaders of the community. These captains tended to be the best village hunters: shrewd judges of ice and men, affluent enough to support a crew and a camp, passing down their equipment and know-how to generations of whalers. In general the captains embraced the opportunities of the oil age—as long as the oil was drilled on land, away from the marine hunting grounds.

When oil companies made efforts to drill in the Beaufort Sea, the captains' association, along with the North Slope Borough, raised alarms about the intrusion of industrial traffic and noise in the migratory corridors of the whales and seals. Above all they feared an uncontrolled oil spill in an icebound ocean, far from clean-up reinforcements. In 2008, Shell Oil bid heavily for federal leases in the Chukchi Sea, and a series of clamorous hearings began on the North Slope. Feelings were particularly strong in Point Hope, which had an unusual history of activism: in the early 1960s the village stopped a plan by government scientists to use "peaceful" nuclear weapons to blast a harbor out of a nearby valley. "There were some very harsh words said about oil companies at meetings here," Oomittuk said. Villagers invoked memories of the starvation that followed the Yankee whalers. Caroline Cannon, an activist who led delegations to Washington, D.C., said at the time, "It feels as if the government and industry want us to forget who we are . . . as if they hope we will either give up or die fighting. We are not giving up." The North Slope Borough and the whaling captains sued to stop the first federal permits, and Shell was forced to retreat. To succeed, the company realized, it would need to find allies among the Eskimos.

A towering wooden fence, 15 feet high and a half mile long, runs across the north side of Point Hope, built by the borough to protect against winds that descend from the North Pole. Before the fence was built, Oomittuk's work as fire chief included shoveling houses out of drifts, sometimes relying on their stovepipes to find them. When I returned to Point Hope last March, I walked along the fence, freshly buried in snow, on the way to Oomittuk's house for a dinner of raw whale meat and caribou stew. The sky was blue and the air calm, but ominous drifts tapered to the south of every house. Soon the wind came, and the next morning Oomittuk's house was frigid: the stove had run out of oil. The wind chill was 38 below, according to my phone. Dishes on shelves on the north wall rattled with each gust. Oomittuk went out in the storm and slipped a two-by-four under one end of the fuel tank to get the oil flowing again, then sent his son off with buckets in search of borough-subsidized fossil fuel.

Oomittuk served as Point Hope's mayor for 10 years, and in those days he opposed offshore oil development. "It was time for us to take care of the animals," he said. "They've taken care of us since time immemorial." But as a steward of traditional culture, he was conscious of the Inupiat principle of *paaqlaktautaiññiq*—the avoidance of community conflict—and he saw that power on the North Slope was shifting. The borough had the money. Tribal councils were losing influence; people had stopped coming to public meetings unless door prizes were offered. And prominent whaling captains had become leaders in business—especially in the land-claims native corporation, the Arctic Slope Regional Corporation.

Starting out in support services for the oil fields that spidered across the region, ASRC had expanded into construction, refining, oil leasing, and government contracting, growing into the largest company in Alaska, with 10,000 employees and gross revenues of $2.5 billion. Because its shares can't be traded publicly, ASRC is subject to few disclosure requirements and can seem opaque. The Alaska Supreme Court recently ruled that it had been unreasonably secretive about executive compensation. But big dividends have tended to quell concerns; in 2013 the Inupiat shareholders—there are 12,000, many of whom live outside the region—received an average of $10,000 each.

The corporation's president, Rex Rock, is a prominent whaler from Point Hope; last year he was the captain of Tariek Oviuk's crew. Rock told me it is not a coincidence that many top ASRC officials are whaling captains. "It's the community's whale," he said. "The captain and crew each know their roles. You work as one to go out and provide for the community. We've taken that into the business world." At the headquarters, a three-story building near the ocean in Barrow, a whaling skin boat provides the center support for a glass-topped boardroom table. But the hunt for profit tells the executives which way to steer: though the company runs television ads of squinting Eskimo hunters declaring "I am Inupiaq," it also took the oil industry's side in a controversial state referendum over oil taxes.

As offshore drilling became a real prospect, ASRC and the industry pressed the argument that whaling and oil could thrive side by side. Shell sponsored boroughwide projects and village feasts and agreed to seasonal drilling restrictions that pleased hunters. Meanwhile ASRC sought to neutralize village opposition. In March, I stopped by a rundown former schoolhouse in Point Hope where a man named Sayers Tuzroyluk was waiting for a computer to be installed in a recently remodeled office. Tuzroyluk, 70 and silver-haired, was the president of Voice of the Arctic Inupiat, a new organization funded by ASRC and the North Slope Borough. The idea, he said, was to line up all the region's tribes and corporations and city governments to speak with one pro-development voice. They were frustrated by hearing anti-oil activists represent the Inupiat in the media, he said: "You have more power when you speak as one voice. We don't speak as individuals. We speak as the whole North Slope."

The biggest change had come in 2010, when the North Slope Borough dropped its legal battle and started working with Shell—first tentatively, then with greater enthusiasm. Last summer, when I went to Barrow to ask about the change, I was ushered into the office of Jacob Adams, the borough mayor's top aide. Before taking office, Adams was the former longtime president of ASRC; he had come out of retirement to help run the borough as Shell's ships were sailing into the Chukchi.

The previous mayor, Edward Itta, had also been willing to sit down with Shell, but he complained that ASRC was pushing the borough too far. "The ASRC are in cahoots with industry, and they're not amateurs at PR," he told me last year. "This campaign up here saying 'I am Inupiaq.' Claiming to be the Voice of the Natives. Well, I'm sorry, they're not."

Like Itta, Adams is an eminent whaling captain. An intense, compact man with gray hair and a crisply pressed business shirt, he told me that the local government could win more safety concessions by negotiating with Shell than by fighting in court. Perhaps more to the point, onshore oil production was declining, and a pipeline coming ashore would mean new roads and facilities, new jobs, more property to tax. The Inupiat, he argued, did not want to go back to hauling lake ice for drinking water, cutting up walrus for dog food, waking up in houses at 25 below zero: "We've created, in the past 40 years, an infrastructure that our children are enjoying now. So will our grandchildren."

Once the whaling captains' association followed the borough and dropped its legal opposition, the tribal council of Point Hope stood almost alone as the indigenous lead plaintiff in an environmental lawsuit against Chukchi Sea drilling. But the melting of the Arctic was drawing new international attention to the cause. Opponents from around the world called the Chukchi prospect an "unexploded carbon bomb" better left in the ground. "Kayaktivists"—inflamed by Shell's many mishaps, including a 2012 debacle that ended with a runaway drill rig in the Gulf of Alaska —prepared to protest the Shell vessels passing through the Pacific Northwest. Caroline Cannon, who told Congress that a major oil spill would amount to "genocide," had been awarded the Goldman Environmental Prize, which came with a stipend of $150,000. Robert Redford narrated a short film about her.

In early 2014 a federal appeals court ruled in favor of Point Hope, blocking Shell's drilling plans for the year, but the pressure on the village only grew. ASRC seemed ready to declare a moratorium on *paaqlaktautaiññiq*. In an open letter, an ASRC executive named Richard Glenn accused the tribes of working with outside environmental groups "to close the door on the future of our communities." As if to emphasize the point, ASRC withdrew several hundred thousand dollars' worth of funding for social and environmental programs run by a boroughwide tribal group that supported the lawsuit. Glenn is a disarming spokesman, a trained geologist with a canny sense of how the wider world likes to see Eskimos: impoverished and clinging to noble tradition. "This is not a Western," he told me. "The word *village* has a quaint image that belies the huge dollar cost of these small cities we have built here. This subsistence lifestyle depends on a lot of money." As he sees it, if environmentalists succeed in closing off the Arctic to oil de-velopment, the North Slope Eskimos will become climate-change victims, no less than if the ice melts away. "We're selfish about our region," he said. "If we sacrifice ourselves, if we shut down all the Arctic, someone elsewhere will turn the valve open a little more."

In July 2014 the new harpoon struck. ASRC called a press con-ference in Anchorage to announce a joint venture with Shell, which would grant the small village-based native corporations a royalty interest in Chukchi Sea oil. Point Hope's village corpora-tion, Tikigaq, was in on the deal. It brought in a professional fa-cilitator for a "visioning" session with leaders of the tribe and the city, and all agreed that without money from offshore oil, their community had no clear path forward. Though anti-oil sentiment remained strong—I was approached many times during my visits by people stressing this—a tribal election was called, and the new officers wrote a letter asking ASRC for help with a budget deficit. Caroline Cannon, the Goldman Prize winner, took a job with the borough mayor. In March 2015, Point Hope withdrew from the Shell lawsuit.

Curiously, worries about the warming Arctic had hardly figured in the region's long debate. When I asked Jacob Adams about the prospects of subsistence in a future without ice, his answer showed a mixture of cultural pride and a hunter's bravado. The Inupiat had always struggled with scarcity and change, he said. They would adapt. The animals would adapt too, he predicted. "Nobody knows whether the ice melting is going to threaten any species at all," he said.

In the late 1970s, when Steve Oomittuk was going to high school in Barrow, he had a job caring for caged animals at the Naval Arctic Research Lab, north of town. Some of the work he found troubling: wolves and marmots were stressed, dehydrated, and needle-jabbed as the government searched for metabolic secrets that could be useful on the battlefield. But several decades of Cold War research were winding down, and soon the Quonset-hut labs were turned over to wildlife biologists working for the North Slope Borough. Under local control, research shifted toward protecting animals that are vital to Inupiat subsistence. Now the borough's Department of Wildlife Management, with an annual budget of more than $5 million, has a mission to marry traditional knowledge and scientific methodology.

A story about the department's origins has become a kind of creation myth for the borough itself: How Oil Saved Subsistence. In 1977 the new borough was confronted by an international effort to halt Eskimo whaling, as regulators claimed that the bowheads had not recovered from the commercial slaughter a century ago. On the advice of elders, the borough's biologists, funded by oil taxes, undertook sophisticated acoustic studies, proving that much of the population had gone uncounted by swimming under the ice. A compromise was reached, eventually allowing Alaska's villages a maximum of 67 whale strikes each year. (Regulators were also concerned that too many novice hunting crews, funded by the new oil wealth, were striking and losing whales, a detail sometimes overlooked in the retelling of the story.)

The warming Arctic became a focus for the borough's biologists. They drew on insights of elders like Arnold Brower Sr., the Inupiaq son of a 19th-century Yankee whaler. Brower, who turned down schooling in San Francisco in order to spend his boyhood in a reindeer camp, landed more than three dozen bowheads in his lifetime, making him one of the most successful captains of his day. In 2001 he described to Charles Wohlforth, in *The Whale and the Supercomputer,* how the weather was undermining traditional knowledge. "You could predict to go out there and hunt all day and not think about getting stranded," he said. "But I think we had a crazy type of change." Brower died a few years later, during an unseasonably late freeze-up. As he traveled alone to his fishing camp, at the age of 86, his snowmobile broke through river ice.

In the fall of 2009 two biologists with the U.S. Geological Survey, on a flight south of Barrow, spotted a sandy beach littered with walrus carcasses: 131 dead, most of them young, evidently trampled in a stampede. Traditionally female walrus and their young rested on drifting pack ice over a shallow offshore feeding area in the Chukchi Sea. In the past decade, as the summer ice has disappeared early, they have been forced onto beaches, where herds are easy to panic. More worrying, they now face the possibility of a two-day commute to reach their feeding grounds.

The difficulty, as the borough's biologists point out, is that no one is sure whether this crazy type of change is causing actual harm. Walrus, which range widely and spend much of their time underwater, are notoriously hard to count. A 2006 regional census tallied 129,000 but conceded that the actual number could be between 55,000 and 507,000. With error intervals so wide, the walrus could practically go extinct before a statistical change was detected, Robert Suydam, a biologist who works for the North Slope Borough, told me. "We're not in a good place to predict the future, so we're in a very poor place to do anything about it," he said.

It's possible that increased sunlight and productivity in the ice-free waters have helped the walrus by providing more food, but biologists don't know. Similarly, bowheads seem to be thriving, and humpback whales, harbor porpoises, and salmon are expanding their range. But as the climate grows warmer, good fortune can turn bad. Biologists worry especially about the corrosive effect of carbon entering cold Arctic waters, which could eventually hurt the zooplankton that bowheads travel so far to devour.

Since 1975, on a barrier island near Barrow, the ornithologist George Divoky has tended a pioneer nesting colony of black guillemots. Each year, as the snow melted earlier, the guillemots produced more numerous young. It was a global-warming success story—except that the ice was drawing away from the shore faster every summer, pulling with it the Arctic cod that the guillemots fed their young. Around 2002 reproduction rates began to decline—even before polar bears, marooned on the island by retreating ice, started to ravage the nests.

Tools for addressing such slow-developing problems are limited. In 2011 the federal government, citing the effects of lost sea ice, listed the Pacific walrus as a candidate for protection under the Endangered Species Act. But for regulators looking to preserve wildlife populations, Eskimo hunters offer an easier target than major producers of carbon emissions do. In the Arctic potential limits on subsistence hunting are met with anger and disdain—not least among business leaders, who use them to argue that villagers and climate-change activists are not natural allies. Rex Rock and Jacob Adams both pointedly recalled for me that environmentalists had tried in the 1970s to stop indigenous whaling, as an illustration of why the Inupiat should mistrust "outside entities." In recent years the North Slope Borough and ASRC joined oil-industry groups in lawsuits that opposed protections for bearded seals and polar bears, which could impinge on future oil facilities.

The village hunters adapt where they can, traveling farther across open water to the broken ice where marine mammals can be found. On St. Lawrence Island they couldn't reach the walrus last spring; a charity shipped in frozen halibut for replacement protein. In Barrow, where ice in the spring is growing thin and unreliable, the majority of bowheads are now taken during an open-water hunt in the fall. Starting last September, whaling crews in high-speed aluminum boats harpooned 15 bowheads and towed them back to the sandy beach north of town. Meat and blubber were shared among local families and sent to relatives and friends; crew pennants flew from captains' homes, inviting neighbors to come eat. The Inupiat were adapting, and thriving. But even in the celebrations a note of menace lurked. Several young bowheads had been found dead from attacks by killer whales, ice-averse predators that are expanding their range in the Arctic. The first such killing anyone remembered was just two years ago.

By the time Shell's offshore drilling finally got underway, last summer, the operation had taken on the familiar air of an Alaska gold rush. The cost of the operation was enormous—$7 billion—and reports of the expense encouraged rumors of a big find. Why else would Shell have gambled so much? State officials were hoping to refill the Alaska pipeline, which was down to one-quarter capacity. In Barrow, Jacob Adams envisioned tax revenues providing for his grandchildren. In the villages there was talk of royalties, corporate dividends, and jobs.

Then, at the end of September, Shell delivered a shocking announcement: it had failed to find sufficient oil in its Chukchi Sea exploration well and was withdrawing from drilling in Alaskan waters "for the foreseeable future." ASRC's president, Rex Rock, predicted "a fiscal crisis beyond measure" for local communities. He blamed excessive federal regulation—rules that had been largely intended to prevent oil spills and protect wildlife. Environmentalists welcomed the retreat as a sign of a new era, but industry analysts suggested that Shell's decision had less to do with a post-Paris future of carbon budgets and carbon taxes than with conventional economics: the low price of oil and the high cost and uncertainty of drilling in the Arctic.

The argument will surely continue. In March the Obama administration proposed a new five-year plan for offshore oil leasing

that includes future sales in the Beaufort Sea and the Chukchi
Sea. But diminished expectations of an offshore bonanza are now
drawing attention to a different scenario, in which the Inupiat no
longer struggle to choose between oil and subsistence: instead
they could lose them both.

While I was in Barrow last year, I drove north along a beach
road, which runs past the old naval research labs and ends at a
small landmark in the world of climate science: a yellow clapboard
house on the tundra with a three-story scaffolded tower. In April
2012 the National Oceanic and Atmospheric Administration's Barrow observatory was one of the first places on Earth to record a
monthly average of 400 parts per million of carbon dioxide in the
atmosphere. That average is now typical for the entire planet. At
the Paris climate conference last December, the threshold that
raised concerns was 450.

At the NOAA facility two young technicians led me to a roof-
top platform looking across the tundra, where polar-bear sight-
ings sometimes bring a squad car from the borough police, with
a second car for backup. They described how prevailing winds off
the ocean provide pure readings of the carbon dioxide that drifts
over the pole from Europe. The technicians pointed out the Dob-
son spectrophotometer: a small silver dome standing alone, like a
miniature planetarium, and tracking changes in the earth's ozone
layer. Forty years ago, when scientists first installed a dome there,
the world was awakening to concerns about ozone-erasing chlo-
rofluorocarbons. The chemicals began to be phased out in 1989,
and now, decades later, the stratosphere shows signs of stabilizing.
The dome on the tundra seemed a small shrine to hope.

When I returned to Point Hope in March, however, none of the
optimism and resolve of the Paris agreements had made its way
north. The Arctic had just seen two months of record temperatures, in a winter that researchers were calling "absurdly warm."
The polar ice pack was thinner, and its maximum extent was the
smallest ever measured. Approaching in a small plane, I could see
open water stretching for miles south from the Tikigaq peninsula.
Hunters in the village had already spotted a bowhead, a month
early, and were wondering if they might have to set up their spring
ice camps on the beach.

A cold snap was settling in, however, and a frozen skim had
formed beyond the rough cuticle of shore-fast ice. In town, snowy

streets gleamed like polished marble. I stopped by the school complex to see Steve Oomittuk, who had a new job as shop teacher. He
told me he had turned it into Inupiat shop, making tools like the
unaaq, a staff with a hook and a pick that hunters use to probe the
ice pack for holes. The school, which has 238 students, was undergoing a $41 million renovation, including a new gym that could
seat the entire village on bleachers. This was not just a testament
to the popularity of the Harpooners (and Harpoonerettes). Point
Hope had been moved, at great borough expense, to a beach six
feet above sea level; the gym, on high pilings, will be the safest
refuge if a storm crashes through town.

Oomittuk was preparing to join the whaling this year. He serves
as a kind of referee after a whale is landed, dividing the catch
among crews according to arcane rules that reach back into pre-
history. It's an exciting time, but it reminds him of the things that
are disappearing, things that may not be recoverable when the oil
runs out — not just knowledge about hunting and survival or the
ceremonies passed down by his grandparents but the ice itself.
"When all that money goes away, what's going to happen to this
next generation?" he said. "They say the native people were no-
madic, following the animals. That's not true about the Tikigaq
people. The animals came to us. We knew they were coming, to
give themselves to us. And the animals go with the ice. If the ice
goes away, the animals go away."

During my visit hunters went out regularly on snowmobiles to
watch the sea. If the open water lingered, they might need aluminum powerboats instead of the quiet skin boats they prefer.
"Maybe we're going to have to go farther out into the ocean, take
chances," a whaler named Hanko told me. "There's going to be
a time in our life when we're hunting in T-shirts and tank tops."
Once the north wind abated, however, the *sikuliaq* ice started to
return. Oomittuk called elders on his cell phone to organize an
Eskimo dance that they hoped would bring the hunters favorable
winds.

Full-costume Eskimo drumming and dancing remains popular
in Point Hope for big cultural celebrations. But with these infor-
mal dances, which are intended to seek favor from the forces of
nature, the animist echoes were a little strong for some; Oomittuk
told me that Christian whaling captains tended to stay away. Still,
six drummers turned out, and dozens of dancers of all ages. They

gathered at the city office, a two-story geodesic dome that doubles as a bingo hall. "We believe if you follow these rituals, the animals will always come to us," Oomittuk said as he pulled a drum made of whale-liver membrane from a carrying case.

The lead drummer that night was a small, animated man named Leo Kinneeveauk, a retired whaling captain with the angular face of a seabird. He started every song with what sounded like a wail. The male drummers, and the women sitting behind them, sang in Inupiaq style: a first verse, plaintive, then a second, furious and loud, as the men lashed handheld drums. In jeans and sweatshirts, the dancers took turns on the floor, joyously simulating the motions of hunters and their prey. After two hours, Steve Oomittuk was tired and happy. He walked home in the late-evening light of springtime in the Arctic. He would wait to see, in the weeks to come, if the dancing would bring the ice back.

ELIZABETH KOLBERT

A Song of Ice

FROM *The New Yorker*

I

NOT LONG AGO I attended a memorial service on top of the
Greenland ice sheet for a man I did not know. The service was
an intimate affair, with only four people present. I worried that I
might be regarded as an interloper and thought about stepping
away. But I was clipped onto a rope, and in any case, I wanted to
be there.

The service was for a NASA scientist named Alberto Behar. Behar, who worked at the Jet Propulsion Laboratory in Pasadena,
might be described as a 21st-century explorer. He didn't go to un-
charted places; he sent probes to them. Some of the machines he
built went all the way to Mars; they are orbiting the planet today
or trundling across its surface on the *Curiosity* rover. Other Behar
designs were deployed on Earth, at the poles. In Antarctica, Behar devised a special videocamera to capture the first images ever
taken inside an ice stream. In Greenland he once sent a flock of
rubber ducks hurtling down a mile-long ice shaft known as a moulin. Each duck bore a label offering, in Greenlandic, English, and
Danish, a reward for its return. At least two made it through.

When Behar died, in January 2015—he crashed his single-engine plane onto the streets of Los Angeles—he was at work on
another probe. This one, dubbed a drifter, looked like a toolbox
wearing a life preserver. It was intended to measure the flow of
meltwater streams. These so-called supraglacial rivers are difficult
to approach, since their banks are made of ice. They are often

lined with cracks, and usually they end by plunging down an ice shaft. The drifter would float along like a duck, collecting and transmitting data, so that by the time it reached a moulin and was sucked in, it would have served its purpose.

Behar was collaborating on the drifter project with a team of geographers at UCLA. After his death the team carried on with the project, which itself became a kind of memorial. When the geographers picked a supraglacial river to toss the drifters into, they called it the Rio Behar.

I flew up to the Rio Behar in July with several UCLA graduate students and two drifters. My first glimpse of it was out the helicopter window. Its waters were an impossible shade, a color reserved, in other circumstances, only for Popsicles. That fantastic blue was set against a pure and hardly less fantastic whiteness. "Greenland!" the artist Rockwell Kent wrote, after being shipwrecked in an ice fjord. "Oh God, how beautiful the world can be!"

An earlier wave of students had already set up a camp. This consisted of one orange cook tent and nine smaller tents, also orange. Beneath the camp the ice extended more than half a mile. Dotting its surface were perfectly round holes, each an inch or two in diameter and about a foot deep. The holes were filled with meltwater. On this half-solid, half-liquid substrate, staking the tents had proved impossible. The one I was assigned was tied to a quartet of fuel canisters. "Don't smoke," someone advised me.

A line of yellow caution tape had been strung about 50 yards from the Behar's edge. Anyone venturing beyond that line, I was instructed, had to be tethered. I borrowed a mountaineering harness, clipped in, and made my way to the bank, where the team's leader, Larry Smith, was conferring with a pair of graduate students. By ice-sheet standards it was a balmy day—around 32 degrees—and Smith was wearing canvas work pants, two plaid shirts, one on top of the other, and a red fleece cap that said AIR GREENLAND.

"Do you hear that?" he asked me. Above the rush of the river there was a roaring sound, like waves crashing against a distant cliff. "That's the moulin."

Eighteen months after the plane crash, Smith still had trouble talking about Behar. He had brought to the river a half-liter bottle of Coke, which he was carrying in a side pocket of his pants. In the field, he told me, Behar had more or less lived on Diet Coke. He apologized for having to substitute the sugared variety.

Smith twisted open the bottle, drank from it, then handed it around. Each of the students took a few swigs. When Smith got it back, he wrote his email address on the label, with the message "If found, please contact." Then he lofted the bottle into the Behar and we all watched it disappear, floating toward the moulin in the icy blue.

People attracted to the Greenland ice sheet tend to be the type to sail up fjords or to fly single-engine planes, which is to say they enjoy danger. I am not that type of person, and yet I keep finding myself drawn back to the ice—to its beauty, to its otherworldliness, to its sheer, ungodly significance.

The ice sheet is a holdover from the last ice age, when mile-high glaciers extended not just across Greenland but over vast stretches of the Northern Hemisphere. In most places—Canada, New England, the Upper Midwest, Scandinavia—the ice melted away about 10,000 years ago. In Greenland it has—so far, at least—persisted. At the top of the sheet there's airy snow, known as firn, that fell last year and the year before and the year before that. Buried beneath is snow that fell when Washington crossed the Delaware, and beneath that, snow from when Hannibal crossed the Alps. The deepest layers, which were laid down long before recorded history, are under enormous pressure, and the firn is compressed into ice. At the very bottom there's snow that fell before the beginning of the last ice age, 115,000 years ago.

The ice sheet is so big—at its center, it's two miles high—that it creates its own weather. Its mass is so great that it deforms the earth, pushing the bedrock several thousand feet into the mantle. Its gravitational tug affects the distribution of the oceans.

In recent years, as global temperatures have risen, the ice sheet has awoken from its postglacial slumber. Melt streams like the Rio Behar have always formed on the ice; they now appear at higher and higher elevations, earlier and earlier in the spring. This year's melt season began so freakishly early, in April, that when the data started to come in, many scientists couldn't believe it. "I had to go check my instruments," one told me. In 2012 melt was recorded at the very top of the ice sheet. The pace of change has surprised even the modelers. Just in the past four years more than a trillion tons of ice have been lost. This is 400 million Olympic swimming pools' worth of water, or enough to fill a single pool the size of New York State to a depth of 23 feet.

An ice cube left on a picnic table will melt in an orderly, predictable fashion. With a glacier the size of Greenland's, the process is a good deal more complicated. There are all sorts of feedback loops, and these loops may in turn spin off loops and subloops. For instance, when water accumulates on the surface of an ice sheet, the reflectivity changes. More sunlight gets absorbed, which results in more melt, which leads to still more absorption, in a cycle that builds on itself. Marco Tedesco, a research professor at Columbia's Lamont-Doherty Earth Observatory, calls this "melting cannibalism." As moulins form at higher elevations, more water is carried from the surface of the ice to the bedrock beneath. This lubricates the base, which in turn speeds the movement of ice toward the ocean. At a certain point these feedback loops become self-sustaining. It is possible that that point has already been reached.

According to the *Encyclopedia of Snow, Ice and Glaciers*, glacial ice "behaves as a non-linear visco-plastic material." To put this differently, ice, like water, flows. For reasons that are not entirely understood, ice flows faster in some parts of the ice sheet than in others. Regions where the flow is particularly swift are known as ice streams.

The East Greenland Ice-Core Project, EGRIP (pronounced "ee-grip") for short, sits atop one of the longest and widest of these streams, the Northeast Greenland Ice Stream, or NEGIS (pronounced "nay-gis"). This past June, I flew up to EGRIP on a ski-equipped C-130 Hercules, which those in the know call a Herc. The Herc had small rockets—jet-assisted takeoff units, or JATOs —mounted below each wing. The JATOs were there in case it got too hot and the runway at EGRIP, which consists entirely of snow, grew sticky.

EGRIP is run by a Danish glaciologist named Dorthe Dahl-Jensen. Dahl-Jensen is a soft-spoken woman with bright blue eyes and an asymmetrical sweep of white hair. She's 58 and has been working on the ice sheet almost every summer for the past 35 years. Initially, as a graduate student at the University of Copenhagen, she'd had to talk her professor, a geophysicist named Willi Dansgaard, into allowing her to come. Dansgaard was against the idea, because the last time he'd brought along a female student the camp's cook had fallen in love with her and stopped cooking. As it happened, on her first trip to the ice sheet Dahl-Jensen fell in love. She and her husband, J. P. Steffensen, also a glaciologist, have four children. During the summer they trade off raising the kids and overseeing operations on the ice.

EGRIP is very much a work in progress. Last year's field season was devoted to hauling equipment from a defunct ice station 275 miles away. This included a whole building, containing a kitchen, a rec room, a bathroom, a dining hall, and an office. The building, which weighs 35 tons, was mounted on skis and dragged behind a tractor equipped with extra-heavy-duty treads.

When I arrived, midway into the 2016 field season, construction at EGRIP was still underway. A network of vaulted tunnels had been created, with floors and walls carved out of snow. These glittered from all angles, like something out of *A Thousand and One Nights*. At the bottom of one tunnel a deep pit had been cut using a chainsaw, and next to the pit a carpenter was erecting a wooden platform. The bricks of ice that had been pulled from the pit had been lugged up to the surface and arranged into what I can only believe is the world's northernmost outdoor bar.

All of this—the tunnels, the pit, the platform—had been fashioned to accommodate an enormous drill, parts of which had traveled with me to EGRIP on the Herc. The point of the project is to send the drill from the top of the ice sheet to the bottom, a distance of more than 8,000 feet. Owing to the way the ice sheet was created, layer upon layer, the drill, as it descends, will in effect be boring through history. (In the case of an ice stream, it is possible to step in more or less the same river not just twice but any number of times.) If all goes as planned, Dahl-Jensen told me, the drilling will be completed in 2020. Meanwhile the ice stream will be moving at the surface, at a rate of around six inches a day, and EGRIP will be moving with it, meaning that the borehole will start to bend. One of the toughest challenges of the project is figuring out how to keep the drill from getting stuck.

The main building at EGRIP—the one that got schlepped across the ice—is a sort of double geodesic dome, with one dome resting on the other like the lid on a casserole. At the very top of it there's

a cupola. The domes and the cupola are covered in rubber sheet-
ing, and to my eye the whole arrangement resembled a big black
time bomb.

My second day at EGRIP, everyone gathered in the double
dome for what was billed as the "first ever" master's thesis defense
on the ice. The chairs in what normally served as the rec room
had been rearranged, classroom-style, and one of Dahl-Jensen's
students, a bearded young man named Kristian Høier, rose to discuss the issue of "surface buckling." Although Høier spoke in English, I couldn't understand most of his presentation, which turned
on details of the equations he'd used in his mathematical model.
He seemed nervous and kept sighing loudly, which I also couldn't
understand, as it was obvious that the first-ever thesis defense on
the ice was going to result in the first-ever pass. When his presentation was over, Dahl-Jensen opened a case of champagne and everyone put on parkas and heavy boots to stand around the outdoor
bar. It was evening but, since the sun never sets in northeastern
Greenland in June, still bright. The snow, flat and unbroken in all
directions, had acquired a bluish tint. Dahl-Jensen offered a toast
to Høier, who seemed intent on getting hammered as quickly as
possible. I left my cup on the bar and went back into the building
to get my camera. By the time I returned, my drink was halfway to
a champagne slushie.

As its name suggests, the NEGIS flows in a northeasterly direction. It has its head, as it were, at the center of Greenland, near the
highest point on the ice sheet. Its mouth empties into the Fram
Strait. There icebergs the size of city blocks split off, or, as geolo-
gists say, calve, and float away. Given enough time, EGRIP, like
some drifting barge, will also reach the Fram and topple in.

All over Greenland, ice streams like the NEGIS are picking up
their pace. In the process they are dumping more and more ice
directly into the oceans. Currently it's estimated that Greenland
is losing about as much ice from calving as it is from melt. One
group of scientists argues that of the two forms of loss, melt is the
more worrisome, as in a warming world it must increase. But the
behavior of ice streams is less well understood, and some scientists
argue that for this very reason increased calving is potentially even
more of a threat.

"The fastest way to get rid of an ice sheet is to throw it into the ocean" is how Sune Olander Rasmussen, the field-office manager for EGRIP, put it to me.

"The ice streams have really, really surprised us," Dahl-Jensen said. "To drill down into an ice stream and see: How does it actually flow? How much is it sliding? How much is it melting at the bottom? I see that as the most important goal of this project."

Once an ice stream starts to accelerate, it may be impossible to stop. "In some cases you have, in theory, this irreversible process," Kerim Nisancioglu, a climate scientist from the University of Bergen who works at EGRIP, told me. "And you set it off and it just goes. It drains.

"This system is huge," Nisancioglu continued, referring to the ice stream we were standing on. "It has a lot of water to drain. So it could keep going for a long time. How far can it go? Will it keep accelerating indefinitely until it runs out of ice? This is unknown." All on its own, the NEGIS has the potential to raise global sea levels by three feet.

The first attempt to drill through the Greenland ice sheet was made in the early 1960s at a United States Army outpost called Camp Century. Some 50 years later, the camp remains far and away the biggest thing ever built on—or really under—the Greenland ice. Camp Century had a bar, a chapel, a barbershop, a movie theater, and a nuclear reactor. All were housed in a network of snow tunnels like those at EGRIP but extending for miles. The ostensible purpose of the base was to promote Arctic science, but in the 1990s an investigation by the Danish government revealed this to be a ruse. What the army had really been up to was developing a new system for storing intercontinental ballistic missiles. Its plan was to install a subglacial railway and shuttle ICBMs around in a Cold War shell game. The code name for the scheme was Project Iceworm.

The drilling at Camp Century was not exactly a secret; still, visitors were not allowed to watch while it was underway. It yielded hundreds of cylinders of ice, each about a yard and a half long and four inches in diameter. These sat around in a freezer in New Hampshire until Willi Dansgaard, Dahl-Jensen's teacher, got hold of them.

Dansgaard, who died in 2011, was an expert on the chemistry

of precipitation. Presented with a sample of rainwater, he could, based on its isotopic composition, determine the temperature at which the precipitation had formed. This method, he realized, could also be applied to snow. Dansgaard was able to read the Camp Century core as a sort of almanac of Greenlandic weather. He could tell how the temperature had changed ice layer by ice layer, which is to say year by year.

Mostly Dansgaard's results confirmed what was already known about climate history. For instance, he observed that Greenland had experienced a cold snap from around the year 1300 to 1800 — the so-called Little Ice Age. He found that for most of the past 10,000 years it had been relatively warm on Greenland, and for tens of thousands of years before that it had been frigid.

But Dansgaard also turned up something totally unexpected. It appeared from his analysis of the Camp Century core that in the midst of the last ice age, temperatures on Greenland had shot up by 15 degrees in 50 years. Then they'd dropped again, almost as abruptly. This had happened not just once but many times.

Everyone, including Dansgaard, was perplexed. A temperature swing of 15 degrees? It was as if New York City had suddenly become Houston or Houston had become Riyadh. Could these violent swings in the data correspond to real events? Or were they some sort of glitch?

Over the next 40 years, five more complete cores were extracted from different parts of the ice sheet. Each time the wild swings showed up. Meanwhile other climate records, including pollen deposits from a lake in Italy, ocean sediments from the Arabian Sea, and stalactites from a cave in China, revealed the same pattern. The temperature swings became known, after Dansgaard and a Swiss colleague, Hans Oeschger, as Dansgaard-Oeschger events. There have been 25 such events in the past 115,000 years.

Ice ages are triggered by small, periodic changes in the earth's orbit that alter the amount of sunlight hitting different parts of the globe at different times of year. The Dansgaard-Oeschger (or D-O) events, which occurred at irregular intervals, have no apparent cause. The best explanation anyone has been able to offer is that the sheer complexity of the climate system renders it unstable — capable of flipping from one state to another,

"It's a great interplay between the glaciers, the atmosphere, the sea ice, and the oceans," Dahl-Jensen told me. We were sitting in

her office, which is in the cupola of the double dome and reach-
able, treehouse-style, via ladder. It was a few hours after the thesis
defense, and the sun was finally dipping toward the horizon.

"But we still struggle to understand how we can get these very
big abrupt changes," she went on. "And I really think that understanding them is one of the most important challenges we face.
Because if we fail to be able to understand them in our past, we
don't have the tools to understand the risk of them in the future."

All the D-O events predate the emergence of civilization, and
this is probably no coincidence. In climatic terms, the past 10,000
years have been exceptionally stable. Go back further than that and
devastating shifts show up again and again. Somehow or other, our
ancestors came through that chaos, but before the invention of
agriculture people traveled light. They never stayed in one place
long enough to develop complex societies and all that followed
—cities, metallurgy, livestock, writing, money. When a D-O event
occurred, bands of hunter-gatherers presumably picked up and
moved on. Either that or they died out.

II

Greenland is the world's largest island, unless you count Australia,
which is usually put in its own category, since it's a continent. The
ice sheet covers about 80 percent of the island, making it one of
the least green places on earth.

"Greenland should be called Iceland and Iceland should be
called Greenland," Inuuteq Holm Olsen, Greenland's representative to the United States, told me with a shrug of irritation. "You
don't know how many times I've heard that." If Greenland were its
own country, it would be the biggest nation in Europe, although,
geologically speaking, it's part of North America. The ice-free territory alone—some 170,000 square miles—is larger than Germany.
As it is, the island is ruled by the Kingdom of Denmark, and Olsen
occupies an office in the basement of the Danish embassy in Washington, D.C. Like most Greenlanders, he's of Inuit descent.

For as long as they could, the Danes kept Greenland under a
sort of reverse quarantine: the goal was not to keep residents in
but everybody else out. Foreigners wishing to visit had to apply to
Copenhagen for approval; the difficulties of obtaining permission,

Rockwell Kent complained in 1930, were "serious and many." (At that point there was no such thing as private property on the island, and indeed even today, in keeping with Inuit tradition, all land is held in common.) According to the Danes, the arrangement was maintained for the good of the Greenlanders, to guard them against the "destructive trends" of modern life. As late as 1940, many families still lived in turf houses and lit their homes with seal-oil lamps.

During the Second World War, Denmark was occupied by the Nazis, and the United States built several airbases on Greenland. By the time the conflict was over, Greenlanders had seen too much of modern life, destructive or otherwise, to go back. What followed was what one Danish chronicler has described as "a social quantum leap unmatched in depth, extent and pace anywhere in the world."

Today Greenland has 56,000 residents, 12,000 Internet connections, 50 farms, and, by American standards, no trees. (The native dwarf willows top out at about a foot.) One Greenlander I met, who'd recently left the island for the first time to attend a meeting in upstate New York, told me that his favorite part of the trip had been the noise of the wind sighing through the leaves.

"I love that sound," he said. *"Shoosh, shoosh."*

There are few roads in Greenland—to get from one town to another you have to take a boat or fly—and, aside from fish-processing plants, little industry. A block grant of $535 million, sent every year by the Danes, constitutes nearly a third of the island's GDP. In a measured, Scandinavian sort of way, relations between the grantor and the grantee are tense.

In 2008, Greenlanders voted overwhelmingly in favor of moving toward independence. Under what's known as the self-rule agreement, which was approved in Copenhagen and in the Greenlandic capital of Nuuk, Greenland gained the right to negotiate some of its own foreign agreements—hence Olsen's basement office. Greenlandic, an Inuit dialect, became the island's official language, and the size of the annual grant from Copenhagen was capped.

Greenland celebrates its version of July Fourth on June 21. This past June, in an effort to demonstrate solidarity, the Danish government instructed its agencies and embassies to raise the Greenlandic flag. A half-red, half-white circle on a half-white, half-red background, the flag is supposed to represent the ice sheet over the ocean, with the sun sinking beneath the waves. Many Danish agencies complied with the directive but, awkwardly enough, flew the flag upside down.

"We have a lot of postcolonial problems," Niviaq Korneliussen, a 26-year-old woman who may be Greenland's most widely read novelist, told me. "We have a lot of racism going on from both ends. There are a lot of young people who hate Danish people because their parents did. So there's a long way to go for things to get better."

Almost a third of the island's population lives in Nuuk, which is by far Greenland's largest town, and in between trips onto the ice sheet I went for a visit. On my 10-minute taxi ride from the airport, I think I passed through all three of Greenland's stoplights.

Nuuk sits on the southwest coast. It was founded in the early 18th century by a Danish-Norwegian missionary named Hans Egede and for most of its existence was known as Godthåb. When Egede arrived, he discovered that the native people had neither bread nor a word for it, so he translated the line from the Lord's Prayer as "Give us this day our daily seal." Today a giant statue of Egede presides over Nuuk much the way Christ the Redeemer presides over Rio.

My visit to Nuuk coincided with a political conclave hosted by Greenland's largest labor union. Many of the island's elected officials were supposed to be there, so one afternoon I made my way over. The walk took me past a set of 10 identical Soviet-style apartment complexes. These were put up in the 1960s, when the Danes decided to empty many of Greenland's tiny fishing villages and concentrate people in larger towns. In their day the apartments, with electricity and indoor plumbing, seemed the height of modernity; now, surrounded by sleeker, newer buildings, they're considered a slum.

The conclave was being held at a large gym with a vaulted ceiling. Inside, about a hundred people were listening to a panel discussion on the subject "Is Greenland ready for the mineral industry?" Simultaneous translation was being provided from Greenlandic into Danish, from Danish into Greenlandic, and from both languages into English. I picked up a headset, but the English channel kept cutting out, and after a while it occurred to me that I was probably the only person trying to listen to it. Tables had been set up around the perimeter of the gym; from them the island's political parties were dispensing sweets, pamphlets, and swag. Groups of impossibly cute children were roaming from one table to the next, grabbing as many balloons and cookies as they could. I struck up a conversation with a man named Per Rosing-Petersen, who was staffing the table for a party called Partii Naleraq. (Almost all Greenlanders nowadays have Danish names, and owing to hundreds of years of intermarriage, many also have blue eyes.) It turned out that Rosing-Petersen was a member of the Greenlandic parliament. Partii Naleraq's offerings included orange plastic bracelets that said *"Tassa asu! Naalagaafinngorta!"* which he translated as "Let's go! Independence!"

"If you look at the businesses in Greenland, 90 percent are owned and managed by Danes," Rosing-Petersen told me. "The Greenlanders are the working class. I call it apartheid—de facto apartheid. We want to change this picture."

Though Greenland's independence movement has nothing directly to do with climate change, indirectly the links are many. For Greenland to break away, it would have to sacrifice the annual grant from Denmark, which would leave a gaping hole in its budget. The island is rich in minerals, and the theory is that these will become easier to get at as winters grow shorter and harbors remain ice-free year-round. Greenland's deposits of rare earth elements are, by some accounts, the largest outside China; the island also has significant deposits of iron, zinc, molybdenum, and gold. In 2014 the Greenlandic government released a plan that called for at least three new mines to be operating within four years. "The mineral resources should—so to speak—be made to work for us," the plan said.

Next to Partii Naleraq's was the table for Siumut, Greenland's ruling party. Manning it was another member of parliament, Jens-Erik Kirkegaard, who, as it happened, had been the minister of industry and mineral resources when the plan was released.

"We haven't had that boon yet," Kirkegaard acknowledged. At the time I visited, the island had no working mines, and the only one under construction—a ruby mine south of Nuuk—was stalled because its Canadian backers had run out of cash. Mostly Kirkegaard blamed the collapse in commodity prices.

"A few years back mineral prices were very high, but then they

declined very hard," he told me. Still, he was optimistic. More melt
off the ice sheet meant more attention for Greenland.

"Climate change does a lot of marketing for us," he said. "It's
easier to attract investment." And as the shipping season grew longer, costs would come down: "Some projects that weren't economical, maybe they will be as conditions change."

Greenland's Institute of Natural Resources, known in Greenlandic as the Pinngortitaleriffik, occupies a stylish wood-and-glass
complex at the edge of Nuuk. The day after the conclave at the
gym I went to the institute to speak to Lene Kielsen Holm, a social anthropologist who studies Greenlanders' perceptions of climate change. Holm does a lot of her work in Qaanaaq, a town in
Greenland's northwest corner that was founded in the early 1950s,
when the U.S. decided to expand one of its airbases—Thule—and
forced most of those living in the area to move out of the way. Qaanaaq, population 630, is one of the few places in Greenland where
people still subsist on what they catch.

"They have always been adapting to a changing environment,"
Holm said of the hunters and fishermen she interviews. "This is
their daily life. If they didn't have this kind of know-how, they
wouldn't survive.

"I think it's part of our culture that we have been living with
changes for a long time," she added.

That Greenlanders are unusually resilient is a view I heard many
times. "Denmark will disappear," Rosing-Petersen told me. "Holland will disappear. But Greenland will still remain. We've been
adapting to living conditions for 5,000 years."

Certainly it's true that life in Greenland is tough. In Qaanaaq,
during the winter months temperatures average around 10 degrees below zero and the sun never appears above the horizon.
"When the long Darkness spreads itself over the country, many
hidden things are revealed, and men's thoughts travel along devi-
ous paths," a west Greenlander told the explorer Knud Rasmussen
sometime around 1904.

But the record of human habitation of Greenland testifies to
more than human resourcefulness. Depending on how you count,
Greenland has been a graveyard for four, five, or even six societies.

The first people to migrate to Greenland are known as the Independence I. This group made its way to the island, probably

from Canada, about 4,500 years ago and settled in a particularly inhospitable territory some 400 miles northeast of where EGRIP sits today. The *Atlas of the North American Indian* notes that the Independence I people "lacked two elements later Arctic dwellers would consider essential: adequate clothing and reliable fuel for fire in a treeless landscape." Somehow they managed to eke out a living for almost a millennium. Then they disappeared.

The Independence I people were followed by a group called Independence II, which also vanished. Meanwhile people known as the Saqqaq arrived in western Greenland. They lasted almost 2,000 years and were replaced by what archaeologists call the Dorset. Recent DNA analysis of their remains suggests that both the Saqqaq and the Dorset died off without descendants. From around the time of the birth of Christ to around the time of Charlemagne, Greenland was, it appears, uninhabited.

In the late 10th century the island was repopulated, this time from the east, by a contingent of Norse led by Erik the Red. It's debated whether Erik called the place Greenland because at that time it really was greener or because he thought it would be good PR. The Norse established two main colonies: the Western Settlement, which was not far from present-day Nuuk, and the Eastern Settlement, which was actually in the south. The settlements prospered and grew until something went terribly wrong. When Hans Egede set out for Greenland in 1721, he was hoping to bring Protestantism to the Norse, who, he worried, had missed out on the Reformation. But all that was left of the settlements was ruins.

Archaeologists have since determined that the Western Settlement failed around the year 1400 and the Eastern Settlement a few decades later. In climatological terms, this timing is suggestive. The Europeans arrived in Greenland during the so-called Medieval Warm Period, and they vanished not long after the onset of the Little Ice Age.

Still, archaeologists have sought alternative explanations for their disappearance. It's been hypothesized that the Norse were overpowered by the Inuit, who arrived in Greenland, also from Canada, sometime around A.D. 1200. Or that they were done in by a drop in the value of walrus ivory. In *Collapse* (2005), Jared Diamond attributes their demise to an oddly self-punishing cultural conservatism. The European settlers had brought with them cattle and sheep. According to Diamond, they continued to rely on their livestock even though they would have been a lot better off copying the Inuit and adopting a marine-based diet.

"The Norse starved in the presence of abundant unutilized food resources," he writes. But according to more recent research, based on the isotopic composition of Norse bones, the Europeans did ditch their cows. By the time the Norse vanished, at least half their calories were coming from seal meat.

"If anything, they might have become bored with eating seals" is how Niels Lynnerup, of the University of Copenhagen, one of the scientists who led the research, put it.

"It's one of those things where, wow, you realize you can be resilient, you can be adaptive, you can be clever, and you can still all be extinct," Thomas McGovern, a professor of archaeology at Hunter College who has studied the Norse for 35 years, told me.

As Greenland warms, the record of the Norse settlements, along with any clues that it might yield, is being erased. "Back in the old days these sites were frozen most of the year," McGovern continued. "When I was visiting south Greenland in the 1980s, I was able to jump down in trenches guys had left open from the '50s and '60s, and sticking out the sides you could see hair, feathers, wool, and incredibly well-preserved animal bones." A graduate student of McGovern's who started working in Greenland in 2005 found at the same sites mostly decomposing mush.

"We're losing everything," McGovern said. "Basically, we have the equivalent of the Library of Alexandria in the ground, and it's on fire."

III

The town of Ilulissat sits 350 miles north of Nuuk, above the Arctic Circle. It's home to one of Greenland's richest archaeological sites—a stretch of springy tundra that was inhabited first by the Saqqaq, then by the Dorset, and finally by the Inuit. Near the abandoned settlement is a bare stone ledge overhanging a fjord. Elderly Greenlanders used to jump from the ledge to avoid becoming a burden to their families, or so the story goes. The day I went to stand on the ledge, several Danish tourists were taking photos and batting away mosquitoes. Instead of jumping, we had come to admire the view.

Rising from the fjord in front of us was a vast, improbable collection of icebergs. These were jammed together as in a frozen
metropolis. Towers of ice leaned against arches of ice, which
pressed into palaces of ice. Some of the icebergs had smaller
icebergs perched on top of them, like minarets. There were ice
pyramids and what looked to me like an ice cathedral. The city
of ice stretched on for miles. It was all a dazzling white except for
pools of meltwater—that fantastic shade of Popsicle blue. Nothing
moved, and apart from the droning of the mosquitoes, the only
sound was the patter of water running off the bergs.

The suicide ledge is a good place to go to feel small—presumably that's why it was chosen. Standing at its edge, I could imagine
how the Saqqaq and the Dorset were awed by the inhuman beauty.
But today even sublimity has been superseded.

The city of ice is the product of the Jakobshavn ice stream. Like
the NEGIS, the Jakobshavn originates in central Greenland, only
it flows in the opposite direction and into a long fjord. Where the
ice meets the water, there's a calving front, and it's here that the
ice arches and ice castles take form. These float down the fjord
toward Ilulissat. (The town's name is Greenlandic for "icebergs.")
They would continue on out to sea, except that they're blocked by
a submarine ridge—a moraine—composed of rocky debris left behind when the ice sheet shrank at the end of the last ice age. The
biggest icebergs become lodged on the moraine and the smaller
ones crowd in behind, as in a monumental traffic jam. The very
largest, which weigh upward of 100 million tons, can hang around
for years before slimming down enough to float free. (It is believed that one of these liberated giants from Ilulissat was the iceberg that sank the *Titanic*.)

Eight thousand years ago the Jakobshavn filled the fjord completely, all the way to the moraine. By the mid-19th century, when
the first observations were recorded, the position of the calving
front had shifted inland by about 10 miles. Over the next 150 years
the front's position shifted again, by another 12 miles.

Then, suddenly, in the late 1990s, the Jakobshavn's stately retreat turned into a rout. Between 2001 and 2006 the calving front
withdrew nine miles. Just in the past 15 years it has given up more
ground than it did in the previous century. The fjord extends for
at least another 40 miles and deepens as it moves inland. At this

point there doesn't seem to be anything to prevent the calving front from withdrawing the entire way.

"It appears now that the retreat cannot be stopped," David Holland, a professor at NYU who studies the Jakobshavn using seals equipped with electronic sensors, told me. (When the seals surface after a dive, the sensors transmit data about conditions in the fjord.)

Meanwhile, as the calving front has receded, the ice stream has sped up. This appears to be the result of yet another feedback loop. Since the '90s the Jakobshavn has nearly tripled its pace. In the summer of 2012 it set what's believed to be an ice-stream record by flowing at the distinctly unglacial rate of 150 feet per day, or more than six feet an hour. The Jakobshavn's catchment area is smaller than the NEGIS's; still, there's enough ice in it to raise global sea levels by two feet.

A lot of Ilulissat is given over to dogs. They have their own neighborhoods—large expanses of dust and rock, where they live chained up around industrial-size vats of water. In my walks around town I encountered three dog settlements that spread over several acres, and behind my hotel there was a small satellite encampment. In the endless summer sun, the dogs looked stricken. They lay around, panting under their thick coats. Occasionally one group would start to bay and then the rest would take up the cry, so that the whole town seemed to be howling.

Ilulissat's dogs are all the same kind, a particularly cold-hardy breed of husky, which the Inuit brought with them when they migrated to Greenland. To maintain the purity of the breed, no other type of dog is allowed north of the Arctic Circle.

The huskies used to be central to Greenlandic life. "Give me dogs, give me snow, and you can keep the rest," Knud Rasmussen, the explorer, who was born in Ilulissat in 1879, supposedly once said. As recently as 1995, Ilulissat, a town of some 4,600 people, was home to more than 8,000 dogs. In the past 20 years the canine population has crashed. Now there are only about 2,000 dogs. This too is an index of global warming.

Ole Dorph, Ilulissat's mayor, works out of a corner office in the town's surprisingly sprawling city hall. He's 61, with a craggy face and rectangular glasses. Dorph grew up in Ilulissat, and he told me that when he was a child, every year the town was iced in from November to April. During those months residents used their dog sleds to go fishing and seal hunting.

"In the old days you could take your sled and go to Disko Island," Dorph said. The island, the largest in Greenland outside of Greenland itself, lies about 30 miles west of Ilulissat, across Disko Bay.

Since no supply ships could get into Ilulissat's harbor, for six months a year residents had to live off whatever provisions the stores had laid in, plus whatever they caught. When the ice broke up in the spring and the first ship arrived, "everyone was very happy," Dorph recalled. "We could buy new apples." To announce the boat's approach, the town would "shoot off a cannon three times — *bang, bang, bang.*"

Then, in the '90s, the bay started to freeze later and later, until finally it didn't freeze at all. "The last time we had ice we could use was in 1997," Dorph told me.

The loss of ice cover from Disko Bay is part of the general decline in Arctic sea ice — a decline that's been so precipitous it now seems likely there will be open water at the North Pole in summer within the next few decades. Since sea ice reflects the sun's radiation and open water absorbs it, the loss has enormous implications for the planet as a whole. (Sea ice doesn't contribute to sea-level rise, because it floats, displacing an equivalent amount of water.) Locally, in Ilulissat, the most obvious impact has been on transportation. Once the bay stopped freezing, supply ships could arrive in January, and sleds became obsolete. Dogs no longer seemed worth the seal meat it took to feed them. Many were euthanized. Those that remain are used mostly for sport.

Dorph told me that people in Ilulissat were "sad because our dogs are going down," but that this unhappiness was more or less balanced by the benefits of open water. Ilulissat's major source of income is halibut, and its small harbor, which sits on the opposite side of town from the fjord, is crowded with fishing boats.

"The fishermen, they can take their boats out in winter," Dorph said. "They feel it's okay. The price of fish is going up, so the fishermen, they have good days." I was reminded of what I'd heard in Nuuk — that climate change, while regrettable in many ways, was for Greenlanders filled with economic promise. I asked Dorph, a member of the ruling Siumut Party, about independence.

"I hope it will happen in maybe 10 or 20 years," he said. "It's our key to growing up."

One evening while I was staying in Ilulissat, I hired a boat to go up the coast. The owner, who was also the captain, was a Dane named Anders Lykke Laursen. He met me at the harbor wearing a pair of yellow-tinted sunglasses, which, he explained, would help him spot dangerous chunks of floating ice. The boat, he went on to assure me, had a double hull and met all the standards the Danish Maritime Authority had laid out for operating in the Arctic. If we did hit some ice, he advised, "it's going to sound bad—but don't worry."

About 10 miles north of Ilulissat we passed the tiny town of Oqaatsut, a collection of bright-painted houses hugging the rocks. (*Oqaatsut* is Greenlandic for "cormorants.") From the boat not a soul was visible, but when I looked it up later in the phone book— there's one edition of the white pages for all of Greenland, and it's about an eighth of an inch thick—I found that Oqaatsut had 18 listed numbers. We motored on, dodging refrigerator-size blocks of floating ice as well as several massive icebergs that had broken free from the moraine. Beyond Oqaatsut the coast rose up. A thin waterfall hundreds of feet high twisted off the rocks. Almost anywhere else in the world the falls would have been a major tourist attraction; in the great emptiness of west-central Greenland, it didn't even have a name.

Finally, after about three hours, we came within sight of our destination, a rock-strewn cove. It also had no name; its coordinates— 69.868245N by 50.317827W—had been sent to me by Eric Rignot, a glaciologist from the University of California, Irvine. The cove was shallow, so we paddled ashore in a rubber dinghy, pushing ice chunks out of the way with the oars.

Rignot, who grew up in France, studies both Greenland's ice sheet and Antarctica's. Two years ago he published a paper arguing that a key section of the West Antarctic ice sheet, the Amundsen Sea sector, had gone into "irreversible retreat." The Amundsen Sea sector contains more than 200,000 cubic miles of ice, meaning that if Rignot's analysis is correct, it will inevitably raise global sea levels by four feet.

"This Is What a Holy Shit Moment for Global Warming Looks Like," *Mother Jones* declared when the paper was released.

Rignot and three of his students had set up camp on a steep hill just beyond the beach—a cluster of pup tents facing a glacier-filled fjord. In the slanted sunlight—it was about 9 p.m.—the glacier, known as Kangilernata, seemed to be glowing. Its calving front, a 130-foot vertical wall of ice, appeared upside down in the milky-blue waters of the fjord. Behind it, ice stretched to the horizon. Again I was hit, and vaguely sickened, by Greenland's inhuman scale.

Rignot and his students were monitoring Kangilernata's movements with a portable radar set, which resembled a rotating badminton net. "We measure changes in the condition of the glacier within millimeters," Rignot told me. "It's like making a movie of the flow." But even without sophisticated equipment the glacier's retreat was apparent. Rignot pointed to a 50-foot-wide band of gray along the walls of the fjord. This showed how much Kangilernata's height had fallen. Coal-black moraines marked the retreat of its calving front. In the past 15 years the front has pulled back two miles.

Kangilernata is what's known as a marine-terminating glacier. So is Jakobshavn, and so too are most of the glaciers in West Antarctica. This means that they have one foot in the water and, as the world warms, are melting from the bottom as well as from the top. NASA is so concerned about this effect that it has launched a research project called, suggestively, Oceans Melting Greenland, or OMG. (Rignot is one of the principal investigators on the project.)

At Kangilernata the team was measuring the water temperatures at the base of the calving front every other day. This involved taking a Zodiac into the fjord, dangling some instruments over the side, and hoping the boat wouldn't be swamped by falling ice.

"What concerns me the most is that this is the kind of experiment we can only do once," Rignot said. "A lot of people don't realize that. If we start opening the floodgates on some of these glaciers, even if we stop our emissions, even if we go back to a better climate, the damage is going to be done. There's no red button to stop this."

I first visited the Greenland ice sheet in the summer of 2001. At that time vivid illustrations of climate change were hard to come by. Now they're everywhere—in the flooded streets of Florida and South Carolina, in the beetle-infested forests of Colorado and

Montana, in the too-warm waters of the mid-Atlantic and the Great Lakes and the Gulf of Mexico, in the mounds of dead mussels that washed up this summer on the coast of Long Island and the piles of dead fish that coated the banks of the Yellowstone River.

But the problem with global warming—and the reason it continues to resist illustration, even as the streets flood and the forests die and the mussels rot on the shores—is that experience is an inadequate guide to what's going on. The climate operates on a time delay. When carbon dioxide is added to the atmosphere, it takes decades—in a technical sense, millennia—for the earth to equilibrate. This summer's fish kill was a product of warming that had become inevitable 20 or 30 years ago, and the warming that's being locked in today won't be fully felt until today's toddlers reach middle age. In effect, we are living in the climate of the past, but already we've determined the climate's future.

Global warming's backloaded temporality makes all the warnings—from scientists, government agencies, and especially journalists—seem hysterical, Cassandra-like—*Ototototoi!*—even when they are understated. Once feedbacks take over, the climate can change quickly, and it can change radically. At the end of the last ice age, during an event known as meltwater pulse 1A, sea levels rose at the rate of more than a foot a decade. It's likely that the "floodgates" are already open, and that large sections of Greenland and Antarctica are fated to melt. It's just the ice in front of us that's still frozen.

On my last day in Ilulissat, I decided that since I might not be coming back, I ought to go see the ice city again. My hike took me through one of the dusty dog encampments and by the town's old heliport, where, to help boost tourism, a Danish philanthropy is planning to erect a viewing platform overlooking the fjord. (The platform "will provide a front-row seat for the melting ice sheet," the head of the philanthropy said in June, when the winning design was announced.) The ice city didn't appear to have changed much, and I recognized some of the same arches and castles I'd seen earlier. It was a cloudless morning, and again, apart from the mosquitoes, nothing was moving. I'd brought along a notebook and started to make a list of the shapes before me. One iceberg reminded me of an airplane hangar, another of the Guggenheim Museum. There was a sphinx, a pagoda, and a battleship; a barn, a silo, and the Sydney Opera House.

To get back to town I followed a different route. This one took me past Ilulissat's cemetery. The plots were marked with white wooden crosses and heaped with bright-colored plastic flowers. It was a lovely and oddly cheerful sight, the graveyard overlooking the ice.

ADRIAN GLICK KUDLER

Something Uneasy in the Los Angeles Air

FROM *Curbed*

PEOPLE WHO DON'T know any better like to say Los Angeles has no seasons, but that isn't true; it has five overlapping seasons: the winter rainy season, spring, gloomy early summer (also known as jacaranda season), miserably hot late summer, which lasts through October, and Santa Ana season. For non-Angelenos, the most LA season is that brief spring, when the days are 72 degrees and sunny. But for Angelenos, who have a far more intimate relationship with both nature and apocalypse than the 72-degrees-and-sunny crowd will ever allow, the most Los Angeles season is Santa Ana season.

The mythology around the Santa Ana winds is potent enough that "Santa Ana winds in popular culture" has its own robust Wikipedia page, and they appear everywhere from Steely Dan's "Babylon Sisters" to Bret Easton Ellis's *Less Than Zero* to a season-four episode of *Beverly Hills, 90210.* But the best-known and most cited appearances are in the opening to Raymond Chandler's story "Red Wind":

> There was a desert wind blowing that night. It was one of those hot dry Santa Anas that come down through the mountain passes and curl your hair and make your nerves jump and your skin itch. On nights like that every booze party ends in a fight. Meek little wives feel the edge of the carving knife and study their husbands' necks. Anything can happen. You can even get a full glass of beer at a cocktail lounge.

and the first part of Joan Didion's essay "Los Angeles Note-book":

> There is something uneasy in the Los Angeles air this afternoon, some
> unnatural stillness, some tension. What it means is that tonight a Santa
> Ana will begin to blow, a hot wind from the northeast whining down
> through the Cajon and San Gorgonio Passes, blowing up sandstorms
> out along Route 66, drying the hills and the nerves to flash point. For a
> few days now we will see smoke back in the canyons, and hear sirens in
> the night. I have neither heard nor read that a Santa Ana is due, but I
> know it, and almost everyone I have seen today knows it too. We know
> it because we feel it. The baby frets. The maid sulks. I rekindle a wan-
> ing argument with the telephone company, then cut my losses and lie
> down, given over to whatever it is in the air. To live with the Santa Ana
> is to accept, consciously or unconsciously, a deeply mechanistic view of
> human behavior.

In a *New York Times* article in 1963, Eugene Burdick (who'd
grown up in LA!) wondered for several pages how on earth California had recently passed New York to become the most populous
state in the country. In 2016 his scene-setting reads like a parody:

> One summer day when a "Santa Ana" wind swept tons of desert dust
> aloft to combine with the smog to give Los Angeles a brown, hazy at-
> mosphere, I visited Muscle Beach at Santa Monica. Sitting on a bench,
> peering through the warm, brown swirling air, were a dozen senior
> citizens watching a group of young men and women go through the
> tortures which produce heavily muscled and almost ridiculously perfect
> physiques.

Like every other common thing in Los Angeles, like everything
else around here that Didion has turned her heavy-lidded eyes to,
the winds have become a part of the story we tell ourselves about
being Angelenos, like earthquakes and irritating development executives at parties, a mysterious force exotic enough to the folks
Back East that they can use it to dismiss us.

Pleasant summer winds form over the Pacific Ocean. Santa Anas
start in the Great Basin, beyond the Sierra Nevadas, in winter,
when the air is cold and the jet stream leaves behind high-pressure
systems, which spin clockwise, cold and dense, until the heavy air
starts to slide down the mountains toward the coast. Lower pressure at the coast helps by sucking that cold air through the moun-

tains toward Southern California. As it cascades down toward the Los Angeles Basin, the air heats up and dries out, and it speeds up as it snakes its way through narrow passes and canyons, barreling out finally in the flats, blowing 110 miles per hour and 110 degrees some days.

Santa Ana season lasts from October to April, but the winds blow just as hard (and sometimes harder) in September and May. Since the air in the Great Basin starts out hotter in those months, the Santa Anas blow hotter in Los Angeles, and they have a lot to do with those miserably hot late summers. "Typically the hottest daytime temperatures along the coast of Southern California have been recorded during Santa Ana winds," Alexander Gershunov, a research meteorologist at the Scripps Institution of Oceanography, says.

Gershunov coauthored a paper published last month in *Geophysical Research Letters* about what he calls "the longest and probably most detailed record of Santa Ana winds available"—from 1948 to 2012. (The lead author was Janin Guzman-Morales, also of Scripps.)

The record reveals patterns in the wind's behavior. They follow "a well-defined diurnal cycle," says the paper, where they're strongest in the morning "and decay to their minimum in the late afternoon." They're more common in El Niño years, when storms off California drop the pressure way low on the coast. They blow most often in December, which is predictable, "because that's when you have the coldest air masses, the longest nights in the High Desert . . . The longest nights and the weakest solar radiation." But some of the strongest winds have blown in the early fall.

That's bad news. In early fall, hillside plants have had all summer to dry out; the Santa Anas suck out any last moisture, and then all it takes is a poorly stamped-out cigarette butt and the hills are on fire, flames fanned by more Santa Anas. Santa Ana fires burn harder, hotter, bigger, faster, and more often than other LA fires, and they burn closer to the city.

Or maybe Los Angeles is lucky and there is no fire on this particular Santa Ana day, but trees are uprooted, power's lost, you wake up to a sickly yellow-pink sky and the dog skidding in frantic circles on the hardwood and the escalating feeling you've forgotten something annoying but important.

*

Can we blame the winds? Raymond Chandler isn't the only one who holds the Santa Anas responsible for bad behavior —they're said to cause migraines, irritability, even suicides and murders.

In the 19th century the winds were thought to be cleansing— an 1886 report from the California State Board of Health called them "health-giving" and informed Californians that after a bout of Santa Ana, "the atmosphere becomes wonderfully clear, pure, and invigorating."

That report also noted an improbable-sounding electricity in the Santa Ana air:

> During the progress of this wind the air is highly electrified. Horses' tails stand out like thick brushes, the hair of the head crackles sharply when rubbed with the hand, and metallic bodies resting on an insulating material, such as dry wood, discharge themselves with visible sparks when a conductor is brought near. In one instance, it is said, the telegraph line between Los Angeles and Tucson, some four hundred and fifty miles in length, was detached from the battery and operated by the earth currents alone.

A man who wrote to the *LA Times* in 1893 to complain about the name of the Santa Anas still had to acknowledge that "it is generally admitted that the winds are beneficial to health, purifying the atmosphere and destroying germs of disease."

But nothing that powerful could possibly be good. By the 1960s the Santa Anas had developed a reputation bad enough to attract a small amount of academic interest—in 1968 a geologist named Willis Miller published his findings that on about two-thirds of Santa Ana days, the homicide count in LA was above average. It's not terribly convincing data, and since then only journalists seem to have looked into the connection. In 2008, *Los Angeles* magazine tallied up a 22 percent increase in domestic abuse reports made to the LAPD during a string of Santa Ana days, and a 30 percent increase in reports to the Santa Ana PD.

Didion wrote that the wind's effects force us to accept a mechanistic view of human behavior. So then what is the mechanism?

The Santa Anas are more or less a type of foehn, an ill wind that blows hot and dry down a mountainside, like the chinook in the far northwest of North America, the khamsin in North Africa, the zonda in the Andes. These hot winds might just be able to blow an electron off an air molecule, creating a precarious but possibly mischievous positive ion.

In the 1950s a bacteriologist named Albert Krueger found that positive ions in the air could drive up the serotonin levels in a mouse's blood and drive it down in the mouse's brain. Serotonin can influence mood, migraines, breathing, and nausea. In 1974 a pharmacologist named Felix Sulman found high serotonin levels in the urine of Israelis who were sensitive to the sharav winds, and prescribed a strong dose of negative ions as the cure.

In 1981 social psychologists Jonathan Charry and Frank B. W. Hawkinshire published research suggesting that

> mood changes . . . were present for most [subjects] when exposed to positive ions, [but] assessment of individual differences in susceptibility was essential for detecting effects on performance and physiological activation. For most [subjects], mood changes induced by ion exposure were characterized by increased tension and irritability.

They also found that when "ion-sensitive" subjects were exposed to positive ions, their skin became less conductive (this is a common psychological gauge) and their reaction times increased.

And in 2000 a group of neurologists published a study that found some migraineurs were more likely to get migraines on days before the chinook blew or on especially windy chinook days. But only two of their subjects got migraines on both types of days, and most got none at all.

So if you've gotten high off ions, get ready for the comedown: a 2013 meta-analysis of ion/mood studies carried out between 1957 and 2012 found "no consistent influence of positive or negative air ionization on anxiety, mood, relaxation, sleep, and personal comfort measures." (It did conclude that negative ions might be able to reduce depression.)

The meek little wife Chandler evokes is a convenient lie. She's just a psychopath or has snorted too much cocaine. That anxious feeling is really a hangover we don't want to admit to ourselves, and who ever knows why the dog does what he does. The science doesn't make a difference; Chandler and Didion and the rest of us just notice late in the afternoon when the air is staticky dry and hot that all day we've been getting the sense that something just beyond our reach has gone sour. It's not the ions. It's just the wind.

*

No one is too eager to tell the truth about the Santa Anas, least of all the Santa Anans of Orange County, whose city is miles away from the Santa Ana Canyon the winds are named for.

Santa Ana fires have burned pretty regularly from at least as far back as 1425, but no one seems to have asked or documented what the Tongva or Chumash called the winds. The earliest Anglos didn't have a name—in a 1943 article in *California Folklore Quarterly*, Terry Stephenson cites Dana Point namesake Richard Henry Dana's recollection of "a violent northeaster" in 1836.

By the end of that century, though, they were the Santa Ana winds. That 1886 California State Board of Health report says the Santa Ana got its name "because it frequently issues from the Santa Ana pass." An angry Santa Anan wrote to the *LA Times* in 1893 that the winds "take the name of Santa Ana by reason of their passage through the Santa Ana mountain cañon" (which was a "gross injustice to Santa Ana and Orange county"). In 1912 the *LA Times* said that "early settlers in this part of Southern California gave the wind its name, because it was alleged to gain access to the region through the Santa Ana Canyon." The 1930s WPA guide for the region says the canyon "gave its name to the hot dry Santa Ana winds that occasionally sweep the southern California coastal counties."

Once he has made clear that "old-timers . . . have always known that the wind got its name because it swept out of the mouth of the Santa Ana canyon," Stephenson documents all the lies about how the winds got their name—a general named Santa Anna was known for his dust-kicking cavalry, there was a notable wind on St. Ann's day (in July!) during the Spanish era, and the one that has stuck:

> The idea was that everybody was mistaken about the name of the wind. It should be called a Santana, which, the Chamber of Commerce was told, was an Indian name for a desert wind . . . Nobody has ever named the tribe that was supposed to have used the name, and nobody has any story as to how away back yonder in the '70's settlers in the Santa Ana Valley managed ingloriously to twist the name into Santa Ana.

By 1967 this story had twisted into this story, in the *LA Times:*

> Others said the Spanish padres translated the Indian term for devil wind into "viente satanas" (wind of Satan).
> Satanas and Santana had been corrupted into Santa Ana, they said.

Santana was and still is widely believed to be the true name of the winds which originated with the Indians.

However, a recognized authority on Indian language says no such word as Santana ever existed.

Like a very boring noir, the Chamber of Commerce seems to be behind so many wrong things we all say about the Santa Ana winds. In 1912 the *LA Times* reported that they had

fathered a movement and campaign of education to get rid of the name Santa Ana as attached to the desert wind that pays occasional visits to parts of Southern California. The directors have passed a resolution asking the newspapers to call the wind a norther or a desert wind, anything so long as it be no longer designated as a Santa Ana wind. The public is called upon to refrain from referring to the wind in letters and conversation as a Santa Ana wind.

Some Orange County businessmen threw a tantrum, and now here we are a century later saying "Devil Winds." On the other hand, as Stephenson writes, "at Santa Ana and everywhere else the wind was still a Santa Ana."

We don't seem to have changed the winds, but we have accidentally helped to make them more dangerous.

Global warming is expected to heat the Great Basin faster than the coast, which should mean less cold air and high pressure to fuel the Santa Anas, but so far that hasn't happened. "There already has been a warming—not as much as we expect in the future—but we don't see any reduction of Santa Ana winds activity in the long record of Santa Ana winds," says Gershunov. He says that the strength with which the winds blow in warmer months like September "tells me the intensity of Santa Ana winds is not controlled just by the temperature of the cold air mass over the Great Basin . . . In the global warming context, it seems that the answer is more complicated."

Actually Gershunov and his coauthors "didn't really see any significant changes in wind frequency or anything else" over 65 years of Santa Anas. Except for one thing: "extreme Santa Ana winds seem to be getting more common, at the expense of run-of-the-mill events," but they don't think that has anything to do with global warming; it seems to correspond instead to the Great

Pacific Climate Shift of the 1970s (which is pretty much what it sounds like, but we'll talk about it another time).

"We don't really understand right now how the Santa Ana winds might change in a warming climate," but scientists have a much better idea of how precipitation will change: there's probably going to be a lot less of it in Southern California. Southern California fire season comes in the fall, later than the rest of the western United States, because of the Santa Anas. But parched vegetation is the fuel, and the longer the dry season lasts into winter, the longer vegetation stays parched, the longer the Santa Ana season has to set it all on fire.

We'll leave our mark before we're done here in the basin, but the Santa Anas were blowing long before Los Angeles began and they'll be blowing long after it's gone. The city they dishevel today isn't the same one Chandler and Didion wrote the myths of so many decades ago—here at the beginning of the 21st century we have different priorities and we're writing new myths. But while we might demolish the freeways and the strip malls, or build towers on every block, the mountains will always rise up in a ring around Los Angeles; the cold, high air will always be pulled down through the canyons, taking on heat, whipping up any palm leaves that are left, unsettling the locals, whatever beasts they may be.

OMAR MOUALLEM

Dark Science

FROM *Hazlitt*

LAST MONTH I found myself on top of a dark mountain attending a "star party."

As much as it sounds like a Beverly Hills soirée or a drug-addled orgy, it is not. This star party is a gathering of a few hundred people at the McDonald Observatory in West Texas to look at the sky. Enormous telescopes were at my disposal, and interpretive astronomers carried high-tech laser pointers that seemingly touched the stars. But otherwise it felt like an ancient ritual.

Peering through a lens at M38—a star cluster too faint for naked eyes that sparkles like diamonds through telescopes—I overheard a girl behind me ask, "What is that?"

"The Milky Way," her mom replied. The girl let out the most awestruck gasp. "See, honey, this is why we get you out of Houston."

City kid. Of course.

Unlike her, I grew up surrounded by countryside, where shortly after sundown anyone could count at least three stars in the sky. By the time Mom tucked me in with stories from the hadith and flicked the lights, the Big Dipper glared at me from outside my window.

I had a generous view of the universe until leaving for college, which I'd forgotten until October 2014, when I returned for a friend's wedding. It was a country party. There was a hog roast. And later, stumbling to town drunk, I found myself in the middle of a field staring at a glistening sky. Had I still believed in him, I'd say it looked like God had sneezed glitter.

The next night I drove home to Edmonton with the aurora borealis by my side. It'd been there all my childhood, but this was the first time I remembered noticing.

Two weeks after the wedding I was on the road again, to the Rockies for Jasper Dark Sky Festival. This is the ultimate star party, with many times more people than were gathered in Texas. The festival started in 2011, shortly after the 11,300-kilometer-square park was designated the world's largest dark-sky preserve by the Royal Astronomical Society of Canada. (Wood Buffalo National Park in northern Alberta, which is four times larger, took the record in 2013.)

The RASC's designations are based on the work of the International Dark-Sky Association, or IDA, which has lobbied dozens of municipalities to adopt dimmer and glare-resistant street lighting, or get rid of it completely, and ultimately improve our views of the night sky. The IDA is headquartered in Tucson, Arizona, but the movement owes much of its success to the RASC for establishing influential light-abatement guidelines and the world's first preserve, in southern Ontario, in 1999. Parks Canada is a partner in the cause, making picturesque Jasper ideal to host such a festival. Outside the park's few towns and lone highway, you basically need a flashlight to navigate the night, or, if you're a hardcore starwatcher, a headlamp with a red bulb that protects your nocturnal vision and only gives as much spotlight as you need to not walk right into a grizzly den.

There are figurative stars at Jasper Dark Sky too, like Colonel Chris Hadfield. I was eager to meet Canada's sweetheart, but as I arrived at the media launch my mind was elsewhere: on my phone and how dead it was, and how my charger was out of reach, and whether I could possibly go three hours without it. In other words, I am exactly the kind of phone-addicted, dead-eyed digital native the festival hopes to net with its slogan, "Power Down, Look Up."

I pumped the home-screen button a few times, but it wouldn't resuscitate, so like a piece of my soul, I left it behind and entered a lodge occupied by marketers, journalists, bloggers, and the guests of honor, members of the RASC. There was a crackling fireplace and warm buffet, but soon I was ushered out the back door and into the cold. The blinds were shut behind me. A group huddled outside around Alister Ling, a meteorologist and club member from Edmonton, who had one of those almighty laser pointers.

But all it could touch were clouds. The sole reason we had come was out of reach.

Time to power up. A fellow member gave Ling an iPad. He held it over his head, and aided by a constellations app, GPS, and brute digital force, he stole back the hidden stars. They glowed red; he tapped a feature that connected the red dots and defined them as ancient Greeks did: gods with simple stories and simpler bodies. It was very unimpressive. Eventually he turned it off and told his own stories in a quiet campfire voice.

Ling reminisced on the last full lunar eclipse, for which he braved frigid weather in a parka, two sleeping bags, and two toques, all to bear witness to a disappearing moon and the purest skylight he'd ever known. "Suddenly I heard a thumping sound," he said. Ling mimicked the noise with his teeth and cheeks. "Then I realized that it was blood rushing through my head."

That he experienced this alone was a point of equal beauty and sadness. Later he would share with me a statistic oft repeated by dark-sky advocates, that two-thirds of the U.S. population can't see the Milky Way anymore. "Artificial light robbed us of our heritage of seeing the sky. It's even worse now." He imitated texting. "A lot of people don't look up."

Even if we could fix the obstruction of artificial light overnight, Ling doubted most people, especially younger ones, would have the patience to make stargazing a hobby. We crave instant gratification, so we're far more likely to stare at the light in our hands than the light in the sky.

It's absurd to think humans should need a protected area or festival to remind them to gaze upward, something that once came to us as naturally as breathing. Given something to see, though, the hardwired urges kick in. During the 1994 Los Angeles earthquakes and subsequent blackout, a nearby observatory was flooded with calls from Angelenos reporting an ominous bright streak looming above. They wondered if the celestial shape was responsible for whatever shook them awake. It was the Milky Way, of course, but it goes to show how primed we are to seek answers in space.

Plains Cree people formed a genesis story around a rupture in the sky, not unlike a wormhole, from which they arrived as spirits before transforming into mortal humans. According to aboriginal educator Wilfred Buck, Cree—"the star people," as Buck calls them—have several names for the Milky Way: *meskinow* (path), *sipi* (river), and *apchak sipi* (spirit river). Like all ancient cultures, his ancestors designated the constellations' names and mythologies in order to package the cosmos into a tidy tale. They were among the first amateur astronomers, a field that today is one of the few areas of interest wherein *amateur* isn't a pejorative but a badge of honor.

Amateurs are responsible for important discoveries about the Milky Way from the last half century, plus some Hubble Telescope equipment that has let us see far beyond it. We have to thank for this movement the so-called International Geophysical Year of 1956 and a Smithsonian initiative called Operation Moonwatch, which for two decades mobilized everyday men, women, and children to help plot the first artificial satellites. What this did for the hobby is encapsulated by this 1958 *Popular Science* headline: "Suddenly, Everybody Wants a Telescope."

The RASC naturally benefited and saw subsequent membership surges following *Apollo 11* and Halley's Comet in 1986. Then, with the amateur-assisted discovery of the Hale-Bopp comet, in 1995, the 4,000-person group soared to 5,000.

That was also the year my parents bought into a growing phenomenon called the Internet, which helped make Hale-Bopp history's most widely observed comet. Observatories' websites tracked it and published daily images; fan sites counted down to its perihelion. In San Diego, Heaven's Gate thought its approach to the sun meant it was prime time for a suicide pact, but in High Prairie, Alberta, 10-year-old me only thought to spend his savings on a cherry-red telescope from the Sears catalogue.

On April 1, 1997, Hale-Bopp slowly scraped the sky with a green tail bright enough for any land mammal to see. But my only memory of my brief brush with amateur astronomy is feeling ripped off by that cheap telescope. Maybe it was just snowing that night, can't remember, but I definitely never touched it again.

I wasn't alone. The RASC's executive director told me membership tanked after Hale-Bopp. But it has bounced back with a renewed interest in astrophotography, thanks to cameras and telescopes being better and cheaper than ever. On their first night out, amateurs capture images orders of magnitude better than professionals did in the 1970s. At the Edmonton chapter, Ling has also witnessed a membership surge, and gender within the historically male-dominated hobby is balancing out. But he said members are of a certain age. "One thing that's really obvious is it's lacking youth." In 1967, 14-year-old Ling's dad lied to the RASC in order for his underage son to qualify for a membership; now it's rare to meet members under 40. "Nerds these days are into gaming."

If amateur astronomy was struggling to be relevant to young people, it didn't show at the family-friendly Dark Sky Festival. The second night was blessed with clear skies, and the disappointment of yesterday turned into hour-long lineups at the entrance gate. I roamed around with a red headlamp and my neck craned to the cosmos, knocking into people equally captivated by the perfectly starry night. We whispered our sorries and bounced between information booths and telescopes manned by interpreters who talked of the value of darkness not just to astronomy but to life itself.

A Parks Canada interpreter and amateur astronomer whose exuberance was fit for children's television struggled to control his volume as he explained to me how artificial light confuses animals' circadian rhythms and effectively leaves nocturnal creatures jet-lagged. That would be a problem for, say, beavers, which are slow, clumsy, and basically defenseless, hence why they work under the cover of darkness despite having crappy eyesight.

I noted the term *scotobiology*—the study of the effects of darkness on biological systems—and wished to learn more about it. I requested a phone call with the interpreter through Parks Canada's communications bureau. It took over four weeks to get approved. Public relations was weirdly nervous and repeatedly asked for my questions in advance. By the time we finally connected, the interpreter's exuberance was overtaken by reticence and scripted explanations to questions I hadn't asked. Apparently embarrassed, he admitted that he was literally reading media lines provided to him. (So much for new prime minister Justin Trudeau's unmuzzling.)

I thanked him for trying and suggested he just email me the interview, since Parks Canada had already conducted it for the both of us. We proceeded to have a friendly chat about his hobby and Cree cosmology, but when I asked again about scotobiology he clammed up. "Scientists believe there is some impact from light on nocturnal animals. Parks Canada is not doing any scientific monitoring on this at the time." He paused. "That is the media line I was given."

In fairness to the interpreter, scotobiology is a very new science. But it's an accepted fact that the last century's combined urban sprawl and electrification is screwing with ecosystems. The research is apparently off-base for Parks Canada, but amateurs and professionals waste no time telling me about whole flocks of birds that've smashed into floodlit smokestacks, and sea turtle hatchlings that are primed to follow moonlight from sand to sea but instead walk inland toward their deaths. "Every nocturnal creature in the world is attuned to light no stronger than the full moon," said Ling. But it might not be limited to nocturnal creatures. "Light increases hormones in chickens and makes them produce more eggs, so is that why girls reach puberty faster?" he wondered. "It may not be all the hormones in the beef." Ling's tendency to blame artificial light for human predispositions isn't exactly junk science. Diabetes, obesity, and certain cancers—preliminary study after preliminary study correlates them with excess light. And yes, one accepted theory for why girls are developing breasts and pubic hair earlier is excessive exposure to light. (Another is absentee fathers, so make what you will of this.)

The dark science has roots in Canada. At a 2003 symposium called "Ecology of the Night," *scotobiology* (from the Greek *skotos,* meaning "darkness") was added to the scientific lexicon. University of Ottawa instructor Robert Dick, who in fact coined the term with biologist Tony Bidwell, remembers attending as an astronomer amongst psychologists, entomologists, and ornithologists. "I went there to talk about the impact on the night sky and they were talking about the impact on biological systems. I was astounded. After a few years of study, I realized darkness had nothing to do with astronomy and everything to do with biology."

A common ground was established. Astronomers would now leverage scientific research to clear the skies, and scientists could leverage astronomers' well-established light-pollution-abatement committees for their causes. After all, health concerns are far more alarming to governments than the argument that, as Ling put it, *you're robbing us of our heritage.* But what's most convincing is cost savings. So across North America, in cities as large as Calgary and Toronto, new bylaws and policies are effectively replacing energy-wasting, halo-casting, upward-facing streetlamps with clearer, softer, downward-facing LED bulbs. A mass relighting of the Western world is well underway.

Dick consulted Ottawa on its 1990s lighting policy before founding his own dark-sky-friendly company for amber lights that don't spoil nighttime vision. He doesn't just sleep with very thick blinds; he takes a self-made amber flashlight to the washroom at night to avoid the shock of white light. To say he's concerned about the effects of artificial light is an understatement. He warns against Big Light's less-than-honest salesmen and compares their denial of adverse health effects to the tobacco industry.

He's particularly worried about blue light, the wavelength ironically emitting from the same anti-light-pollution LED bulbs that cities are adopting en masse, as well as whatever you're probably using to read this. During the day blue light boosts moods and reaction times, but too much of it after sundown can reduce your melatonin, a hormone that sets your internal clock and may be responsible for the aforementioned health problems. As if it wasn't enough that staring into my smartphone for hours every night was distracting me from the night sky, it was now trying to kill me. I suppose I could invest in orange-tinted, Bonoesque glasses that block blue light, like Alister Ling recommended. He also bought a pair for his teenage daughter, but she refuses to wear them. "She's too headstrong to take my advice," he told me. "I don't get it."

It's easy to want to call anti-light-pollution activists Luddites and remind them of the incredible access we now have to outer space via our digital one. After all, it was the instantly gratifying Internet that formed my own interest in cosmology and humanism.

As dust collected on my cheap cherry telescope, I spent countless hours exploring the Web, which was starting to feel as infinite and ever-expanding as the universe itself. That's where I latently discovered Carl Sagan, which led to astronomers Phil Plait and Neil deGrasse Tyson, a pompous detour into Richard Dawkins territory, before I fell in love with Chris Hadfield's Twitter, like just about everyone else.

But for amateur astronomers and dark-sky advocates, who heavily rely on the Web for recruitment and public outreach, this is a supplement, not a substitute, for the real thing—a glittering sky versus a glowing red iPad screen. Virtual Hubble images and viral space videos lunge at me like comets, but all the while my real view of the universe disappears and with it . . . what exactly? Standing under Jasper's skies—or on the Davis Strait mountains or in a field outside my hometown—I was overtaken with a daunting and yet calming sensation that I couldn't quite pinpoint. Ling said it was a sense of my own insignificance, but whatever it was, its absence feels quite significant.

On the last night of the Dark Sky Festival, the press got thirty minutes with Chris Hadfield before his keynote address. I asked him, given that more people now lived in cities than not, was he concerned that humans were losing touch with the sky?

The colonel smiled with a smugness that only someone who's spacewalked could possess. "Not much," he said, "because I've seen the whole world. Everyone who writes about overcrowding lives in a city. It's quite comical. But most of the world is empty. And most of it is dark at night."

MICHELLE NIJHUIS

The Parks of Tomorrow

FROM *National Geographic*

ASSATEAGUE ISLAND NATIONAL SEASHORE, which sits on a 37-mile-long sliver of land just off the coast of Maryland and Virginia, is gradually shuffling west. Over centuries, as hurricanes and nor'easters drive sand from its Atlantic beaches across the island and into its bayside marshes, the entire island is scooting closer to the coast.

"It's neat, isn't it?" says Ishmael Ennis, hunching against a stiff spring wind. "Evolution!" He grins at the beach before him. It's littered with tree stumps, gnarled branches, and chunks of peat the size of seat cushions—the remains of a marsh that once formed the western shore of the island. Later buried by storm-shifted sand, it's now resurfacing to the east, as the island shuffles on.

Ennis, who recently retired after 34 years as maintenance chief at Assateague, has seen his share of storms here. This national seashore, in fact, owes its existence to a nor'easter: In March 1962, when the legendary Ash Wednesday storm plowed into Assateague, it obliterated the nascent resort of Ocean Beach, destroying its road, its first 30 buildings, and its developers' dreams. (Street signs erected for nonexistent streets were left standing in a foot of seawater.) Taking advantage of that setback, conservationists persuaded Congress in 1965 to protect most of the island as part of the National Park System. Today it's the longest undeveloped stretch of barrier island on the mid-Atlantic coast, beloved for its shaggy feral ponies, its unobstructed stargazing, and its quiet ocean vistas—which have always been punctuated, as they are on other barrier islands, by impressive storms.

Scientists expect that as the climate changes, the storms will likely strengthen, sea levels will keep rising, and Assateague's slow westward migration may accelerate. Ennis knows the island well enough to suspect that these changes are underway. Assateague's maintenance crew is already confronting the consequences. On the south end of the island, storms destroyed the parking lots six times in 10 years. The visitors' center was damaged three times. Repair was expensive, and after fistsize chunks of asphalt from old parking lots began to litter the beach, it began to seem worse than futile to Ennis.

A tinkerer by nature—he grew up on a small farm on Maryland's Eastern Shore—he realized the situation called for mechanical creativity. Working with the park's architect, Ennis and his coworkers adapted the toilets, showers, and beach shelters so that they could be moved quickly, ahead of an approaching storm. They experimented with different parking-lot surfaces, finally arriving at a porous surface of loose clamshells—the kind often used on local driveways—that could be repaired easily and, when necessary, bulldozed to a new location. "It was a lot of what we called 'Eastern Shore engineering,'" Ennis says, laughing. "We weren't thinking about climate change. We did it because we had to." He lowers his voice, mock-conspiratorially. "It was *all by accident.*"

Accidental or not, these modest adaptations were the beginning of something broader. The seashore is now one of the first national parks in the country to explicitly address—and accept—the effects of climate change. Under its draft general management plan, the park will not try to fight the inevitable: it will continue to move as the island moves, shifting its structures with the sands. If rising seas and worsening storm surges make it impractical to maintain the state-owned bridge that connects Assateague to the mainland, the plan says, park visitors will just have to take a ferry.

When Congress passed the act creating the National Park Service in the summer of 1916, it instructed the agency to leave park scenery and wildlife "unimpaired for the enjoyment of future generations." The law did not define *unimpaired.* To Stephen Mather, the charismatic borax magnate who served as the first director of the Park Service, it meant simply "undeveloped." Early park managers followed his lead, striving both to protect and to promote sublime vistas.

But the arguments began almost as soon as the agency was born. In September 1916 the prominent California zoologist Joseph Grinnell, writing in the journal *Science,* suggested that the Park Service should protect not just scenery but also the "original balance in plant and animal life." Over the next few decades wildlife biologists inside and outside the agency echoed Grinnell, calling for the parks to remain "unimpaired," in ecological terms. But the public came to the parks for spectacles—volcanoes, waterfalls, trees you can drive a car through—and preserving them remained the agency's primary concern.

In the early 1960s, secretary of the interior Stewart Udall—who would oversee the addition of nearly 50 sites to the National Park System, including Assateague—became concerned about the agency's management of wildlife in the parks. He recruited University of California wildlife biologist Starker Leopold, the son of famed conservationist Aldo Leopold, to chair an independent study.

The Leopold Report proved hugely influential. Like Grinnell, it called on the Park Service to maintain the original "biotic associations" that existed at the time of European settlement. In the decades that followed, the Park Service got more scientific. Park managers began setting controlled fires in forests where natural wildfires had long been suppressed; they reintroduced species that had vanished, such as wolves and bighorn sheep. The focus, though, was less on restoring ecological processes than on recreating static scenes—on making each park, as the Leopold Report recommended, into a "vignette of primitive America." In time that vision took on what Yellowstone historian Paul Schullery describes as an "almost scriptural aura."

And yet, as Leopold himself later acknowledged, it was misleading. The notion of presettlement America as primitive ignored the long impact Native Americans had had on park landscapes, through hunting and setting fires of their own. It ignored the fact that nature itself, left to its own devices, does not tend toward a steady state—landscapes and ecosystems are always being changed by storms or droughts or fires or floods, or even by the interactions of living things. The ecological scenes the Park Service strove to maintain, from a largely imagined past, were in a way just a new version of the spectacles it had always felt bound to deliver to visitors.

"The Park Service has had a tacit agreement with the American public that it's going to keep things looking as they've always looked," says Nate Stephenson, an ecologist who studies forests at Sequoia, Kings Canyon, and Yosemite National Parks. "But time does not stop here."

From the 1980s on, scientists gradually came to accept that a new sort of change was underway. The glaciers in Glacier National Park were shrinking, wildfires in Sequoia were getting larger, and coastal parks were losing ground to rising seas. Shortly after the turn of the century, researchers in Glacier announced that by 2030 even the park's largest glaciers would likely disappear.

In 2003 a group of researchers at the University of California, Berkeley began to retrace the footsteps of Joseph Grinnell. In Yosemite and other California parks, the zoologist had conducted fanatically detailed wildlife surveys, predicting their value would not "be realized until the lapse of many years, possibly a century." When the Berkeley researchers compared their own Yosemite surveys and other data with Grinnell's 90-year-old snapshot, they noticed that the ranges of several small mammals had shifted significantly uphill, toward the ridgeline of the Sierra Nevada. Two other once-common mammals, a chipmunk and a wood rat, were almost extinct in the park. The pattern was clear: climate change had arrived in Yosemite too, and animals were migrating to escape the heat.

For a while the Park Service avoided talking about the subject. To acknowledge the reality of human-caused climate change was a political act, and the Park Service doesn't discuss politics with its visitors. At Glacier the interpretive signs made only a passing reference to rising temperatures. Rangers avoided talk of causes. "We were very constrained," remembers William Tweed, former chief of interpretation at Sequoia and Kings Canyon. "The message we got from above was basically, Don't go into it if you can help it."

The problem, though, ran deeper than transient politics. People had long come to national parks to experience the eternal—to get a glimpse, however deceptive, of nature in its stable, "unimpaired" state. The inconvenient truth of climate change made it more and more difficult for the Park Service to offer that illusion. But no one knew what the national parks should offer instead.

*

When Nate Stephenson was six years old, his parents fitted him with boots and a hand-built wooden pack frame and took him backpacking in Kings Canyon National Park. For most of the 53 years since, Stephenson has been hiking the ancient forests of the Sierra Nevada. "They're the center of my universe," he says. Soon after he graduated from UC Irvine, he packed up his Dodge Dart and fled Southern California for a summer job at Sequoia National Park. Now he's a research ecologist there, studying how the park's forests are changing.

While park managers are often consumed by immediate crises, researchers like Stephenson have the flexibility—and the responsibility—to contemplate the more distant future. In the 1990s this long view became deeply disturbing to him. He had always assumed that the sequoia and foxtail pine stands surrounding him would last far longer than he would, but when he considered the possible effects of rising temperatures and extended drought, he wasn't so sure—he could see the "vignette of primitive America" dissolving into an inaccessible past. The realization threw him into a funk that lasted years.

"I was a firm believer in the mission of the Park Service," Stephenson remembers, "and suddenly I saw that the mission we had was not going to be the same as the mission of the future. We could no longer use the past as a target for restoration—we were entering an era where that was not only impossible but might even be undesirable."

Stephenson began what he calls a "road show," giving presentations to Park Service colleagues about the need for a new mission. Somewhat mischievously, he proposed a thought experiment: What if Sequoia National Park became too hot and dry for its eponymous trees? Should park managers, who are supposed to leave wild nature alone, irrigate sequoias to save them? Should they start planting sequoia seedlings in cooler, wetter climes, even outside park boundaries? Should they do both—or neither?

His audiences squirmed. Leopold had left them no answers.

On a late September day in Sequoia National Park, the sky is clear, blue, and, thanks to a brisk wind, free of smoke from the wildfire burning just over the crest of the Sierra Nevada. Stephenson and his field crew are finishing a season of forest surveys, adding to a decades-long record of forest health. In their lowest-elevation study sites, below the sequoia zone, 16 percent of the trees have died this year, approximately tenfold the usual rate. "It's about what you'd see after a low-grade wildfire," says Stephenson. Weakened by years of drought, many of the low-elevation trees are dying from insect attacks. At higher elevations, in the sequoia stands, several old giants have dropped some of their needles to combat drought stress; a few that were already damaged by fire have died. "It's not 'The sequoias are dying,'" Stephenson emphasizes. "The sequoias are doing relatively well. It's the pines, the firs, the incense cedars—the whole forest is affected."

The current drought may be a preview of the future, but the trouble with climate change—at Sequoia and elsewhere—is that many of its effects are hard to predict. Average temperatures at Sequoia will rise, and snow will give way to rain, but it's not clear whether total precipitation will increase or decrease, or whether the changes will be gradual or abrupt. "We don't know which scenario is going to play out," says Sequoia and Kings Canyon superintendent Woody Smeck. The Park Service can no longer recreate the past, and it can't count on the future. Instead it must prepare for multiple, wildly different futures.

In 2009, Park Service director Jonathan Jarvis assembled a committee of outside experts to reexamine the Leopold Report. The resulting document, "Revisiting Leopold," proposed a new set of goals for the agency. Instead of primitive vignettes, the Park Service would manage for "continuous change that is not yet fully understood." Instead of "ecologic scenes," it would strive to preserve "ecological integrity and cultural and historical authenticity." Instead of static vistas, visitors would get "transformative experiences." Perhaps most important, parks would "form the core of a national conservation land- and seascape." They'd be managed not as islands but as part of a network of protected lands.

The report is not yet official policy. But it's the agency's clearest acknowledgment yet of the changes afoot and the need to manage for them. Exactly what that management looks like isn't certain, and much of it will be worked out park by park, determined by science, politics, and money. Some parks have already gone to great lengths to resist change: Cape Hatteras National Seashore, for instance, spent almost $12 million to move a famous lighthouse a half mile inland. But such dramatic measures are rare and likely to remain so; the Park Service budget today is about what it was in 2008.

Instead many parks are looking to boost their tolerance for change, adapting their own infrastructure and helping their flora and fauna do the same. At Indiana Dunes National Lakeshore, scientists are searching the oak savannas for cooler microclimates into which the Park Service might transport the endangered Karner blue butterfly, which has been all but driven from the park. In Glacier, biologists have already captured bull trout and carried them in backpacks to a higher, cooler lake outside their historic range. The idea is to give the fish a refuge both from climate change and from invasive lake trout.

At Sequoia, Stephenson wants park managers to consider planting sequoia seedlings in a higher, cooler part of the park—to see how the seedlings fare, and also how the public would respond to experimenting with the icons. "We have to start trying things," he says.

At Assateague, while Ennis's successors prepare the parking lots and toilets for change, Liz Davis, the chief of education, is preparing the park's younger visitors. In 25 years at Assateague she has introduced countless school groups to the seashore. When elementary students visit, she takes them to the beach, shapes a model of the island out of sand, and throws a bucket of seawater across it to show how the island shifts. Then she turns the model over to the kids: Where would they put the parking lots and campgrounds? How about the visitors' center? "They get really into it," she says, laughing. "They'll say, No, no, don't put the new ranger station there, it'll get washed away!"

Like the Park Service, visitors must learn to accept that their favorite park might change. "People ask, Will I still be able to enjoy it? Will my kids and grandkids be able to enjoy it?" Davis says. "The answer is yes, they will. They might not enjoy it in the same way, and they might not get here the same way. But they will still be able to enjoy it."

TOM PHILPOTT

How Factory Farms Play Chicken with Antibiotics

FROM *Mother Jones*

THE MASSIVE METAL double doors open and I'm hit with a *whoosh* of warm air. Inside the hatchery, enormous racks are stacked floor to ceiling with brown eggs. The racks shake every few seconds, jostling the eggs to simulate the conditions created by a hen hovering atop a nest. I can hear the distant sound of chirping, and Bruce Stewart-Brown, Perdue's vice president for food safety, leads me down a hall to another room. Here the sound is deafening. Racks are roiling with thousands of adorable yellow chicks looking stunned amid the cracked ruins of their shells. Workers drop the babies into plastic pallets that go onto conveyor belts, where they are inspected for signs of deformity or sickness. The few culls are euthanized, and the birds left in each pallet are plopped on something like a flat colander and gently shaken, forcing their remaining shell debris to fall into a bin below. Now clean and fluffy, the chicks are ready to be stacked into trucks for delivery to nearby farms, where they'll be raised into America's favorite meat.

Not long ago this whole protein assembly line might have been derailed if each egg hadn't been treated with gentamicin, an antibiotic the World Health Organization lists as "essential" to any health-care system, crucial for treating serious human infections like pneumonia, neonatal meningitis, and gangrene. But the eggs at Perdue's Delmarva chicken production farms have never been touched by the drug.

That's extremely uncommon in corporate factory farming. Currently, livestock operations burn through about 70 percent of the "medically important" antibiotics used in the nation—the ones people need when an infection strikes. Microbes that have evolved to withstand antibiotics now sicken 2 million Americans each year and kill 23,000 others—more than homicide. Even though public health authorities from the Food and Drug Administration and the Centers for Disease Control and Prevention have long pointed to the meat industry's reliance on antibiotics as a major culprit in human resistance to the drugs, the FDA has never reined in their use.

I'm in Delmarva, the peninsula composed of pieces of Maryland, Virginia, and Delaware, because it is Big Chicken country —the teeming barns that dot its rural roads churn out nearly 11 million birds per week, almost 7 percent of the nation's poultry. And Perdue, the peninsula's dominant chicken company and the country's fourth largest poultry producer, has set out to show that the meat can be profitably mass-produced without drugs. In 2014 the company eliminated gentamicin from all its hatcheries, the latest stage of a quiet effort started back in 2002 to cut the routine use of antibiotics from nearly its entire production process.

In 1928, Scottish biologist Alexander Fleming discovered a mold-based compound dubbed penicillin that could kill common microbes that cause dangerous infections. But even as they began to revolutionize medicine, antibiotics had a fundamental flaw. While collecting the 1945 Nobel Prize in medicine, Fleming warned that it's "not difficult to make microbes resistant to penicillin in the laboratory by exposing them to concentrations not sufficient to kill them."

When an antibiotic attacks a colony of bacteria, the great bulk of the bacteria die or can no longer reproduce, and the infection is cured. But a few rogue microbes can withstand the assault and pass their hardy genes on to their progeny. If you unleash the same antibiotic on the same bacteria over a long enough time, you'll create a bacterial strain that can thumb its nose at the drug. And it isn't just turbocharged Darwinism that makes our antibiotics so vulnerable. Through a process called "conjugation," genes —including ones that have become resistant to particular antibiotics—can bounce from one microbe to another.

Though we've known this for more than 70 years, doctors have too often treated antibiotics as a sturdy crutch, not a delicate tool to be used sparingly. The CDC estimates up to half of all antibiotics used in U.S. medicine are improperly prescribed, speeding up resistance. But that is nothing compared with how recklessly they've been used in factory farms. When you treat thousands of chickens in a huge enclosed barn with, say, steady doses of tetracycline, you risk generating an *E. coli* bug that can resist the antibiotic you threw at it, and that bug's new superpowers can also jump to a strain of salmonella that happens to be hanging around. Now two nasty pathogens that plague humans have developed tetracycline-resistant strains.

And the worst part is that antibiotic use in factory farms isn't mostly a matter of keeping animals healthy. In 1950 a pharmaceutical company called American Cyanamid—now part of Pfizer—wanted to see if giving chickens vitamin B_{12} made them fatter, so it ran some experiments. The idea seemed to work. But the researchers soon discovered it wasn't the vitamin that had fattened the birds; it was traces of an antibiotic called aureomycin. (B_{12} can be a byproduct of aureomycin production; the vitamin researchers used had come from making the antibiotic.)

This discovery revolutionized meat production. Adding a dash of antibiotics to feed and water rations magically made birds, pigs, and cows grow plumper, saving on feed costs and slashing the time it took to get animals to slaughter. In 1977 the General Accounting Office reported that "the use of antibiotics in animal feeds increased approximately sixfold" between 1960 and 1970. "Almost 100 percent of the chickens and turkeys, about 90 percent of the swine and veal calves, and about 60 percent of the cattle raised in the United States during 1970 received antibiotics in their feed."

In turn, antibiotics helped launch the rapid industrialization of chicken production. In 1950, at the dawn of the antibiotic revolution, 1.6 million U.S. farms were raising about 560 million birds. By 1978 only 31,000 mostly large chicken farms remained, and they were cranking out more than 3 billion birds.

Meanwhile a steady accumulation of scientific research revealed a mounting public health crisis. At the end of the '60s, a scientific committee in the United Kingdom found that using antibiotics in animal feed produced large numbers of resistant bacteria that could be transmitted to people. Similar findings were reported by an FDA task force in 1972, and as a result the agency issued regulations requiring drug manufacturers to prove their agricultural products didn't contribute to resistance. If they couldn't, their approval to sell the drugs would be revoked.

So the Animal Health Institute, a trade group of animal-pharmaceutical manufacturers, contacted Stuart Levy, a young Tufts University researcher who specialized in antibiotic resistance. The group wanted Levy to feed tiny daily doses of antibiotics to chickens and see if the bacteria in their guts developed resistance. The drug companies were convinced the results would "get the FDA off their backs," says Levy.

Levy found a family farm near Boston and experimented on two flocks of chickens. One got feed with small amounts of tetracycline. The other went drug-free.

Within 48 hours, strains of *E. coli* that were resistant to tetracycline started to show up in the manure of the birds fed drugs. Within a week, nearly all the *E. coli* in those birds' manure could resist tetracycline. Within three months, the *E. coli* showed resistance to four additional antibiotics the birds had never been exposed to: sulfonamides, ampicillin, streptomycin, and carbenicillin. Most striking of all, researchers found that *E. coli* resistant to multiple antibiotics was appearing in the feces of the farmers' family members—yet not in a control group of neighbors.

The results, published in the *New England Journal of Medicine*, were so stunningly clear that Levy thought they would prompt the industry to rethink its profligate antibiotic use, or at least inspire the FDA to rein it in. But the industry rebuffed the study it had bankrolled, questioning the validity of the data, Levy says. In 1977 the FDA proposed new rules that would have effectively banned tetracycline and penicillin from animal feed, but the House agriculture appropriations subcommittee, led by agribusiness champion Representative Jamie Whitten (D-Miss.), ordered the FDA to wait, "pending the outcome of further research."

Those proposed 1977 bans remained in limbo for decades, dormant but officially "under consideration"—until 2011, when the FDA finally ditched them and let companies take a voluntary approach to curtailing antibiotic use. Meat producers were given until the end of 2016 to wean themselves from antibiotics. At that point, the agency warned, it would consider banning certain antibiotics if its light-handed approach did not "yield satisfactory results." However, the plan also gave industry a gaping loophole:

while it suggested livestock producers should no longer use hu-
man-relevant antibiotics as growth promoters, it left companies
free to use them to prevent disease.

Unsurprisingly, the industry's appetite for antibiotics has remained voracious. According to the FDA's latest figures, antibiotic
use on U.S. farms surged 13 percent between 2009 and 2014, even
as overall U.S. meat production leveled off. In 2014 livestock operations used 20 million pounds of antibiotics important to humans
—while doctors used about 7 million pounds.

A few years ago I came across a *Consumer Reports* study of bacteria
on supermarket chicken. The magazine had found that Perdue
chicken was far less likely to carry bacteria resistant to more than
one antibiotic than chicken from other producers, including Tyson and Foster Farms. *Consumer Reports* said Perdue's good showing marked the "first time since we began testing chicken that
one major brand has fared significantly better than others across
the board." When I asked Urvashi Rangan, a *Consumer Reports* researcher, why Perdue had done so well, she credited the company's policy against using growth-promoting antibiotics.

As a longtime critic of industrial agriculture, I was surprised to
learn Perdue *had* such a policy. I knew it was a leading supplier of
chicken labeled "no antibiotics ever," but this study showed that
even its conventional chicken was far less likely to carry resistant
bacteria than such chicken from other producers. Could it be that
this one company was tacking against industry norms and taking
the science seriously?

Historically, most birds bound for market not only got antibiotics in their feed but were dosed with drugs before they even
hatched. Bruce Stewart-Brown, who is 59 years old and trim, explains why as we walk through the hatchery. About 40 years ago, a
herpes virus called Marek's disease began to attack chickens, and
vets discovered that vaccinating the chicks while they were still in
their shells could inoculate them for life. But when you penetrate
eggs with a needle loaded with the vaccine, the tiny hole you create opens a door, welcoming bacteria in. To solve this problem,
hatcheries added small amounts of gentamicin to the vaccine to
prevent bacteria from getting a foothold in the bird.

This method was so efficient that decades later the hatchery
ended up being the trickiest place for Perdue to remove antibi-

otics from production. The company gets its eggs from contract breeders, and in the past eggs often arrived covered in bacteria-laden manure. Now Perdue requires its breeders to deliver clean eggs. Perdue also used to mix its Marek's vaccines in the middle of a less-than-pristine hatchery. Today the company mixes the drugs under sterile laboratory conditions and injects clean, antibiotic-free vaccines into clean eggs. It took a while, but by March 2014 the company had banished antibiotics from all 16 of its hatcheries.

Stewart-Brown and I leave the hatchery and head to one of the company's contract farms, where chicks are fattened for slaughter. As we drive through rolling corn and soybean fields, we have a surprisingly blunt conversation. "We already know that we create resistance with the products we use, and we've known that for years," he says. Perdue hasn't studied whether these resistant bacteria leave the farm and endanger the public, via either workers or the chicken sold to consumers. Then he says something remarkable for an agribusiness exec: "We know there's likely some sort of transition."

He's right. The results of Stuart Levy's 1976 study—showing that farmers who come into contact with antibiotic-treated birds quickly pick up drug-resistant bacteria—have been corroborated several times. In a 2013 FDA study of chicken bought nationwide, 60 percent of the salmonella that was detected could resist at least one antibiotic, meaning consumers are one unclean cutting board or unwashed hand away from a nasty bug. Antibiotic-resistant bacteria have also been found in chicken manure, on vegetables grown in fields fertilized with it, and even, according to a 2008 Johns Hopkins study, on the interior surfaces of cars that drive behind trucks hauling chickens to slaughter.

After 20 minutes we pull into a driveway toward two long chicken houses, 40 feet wide and stretching 500 feet back. As with most U.S. chicken facilities, these are run by a quasi-independent farmer who is paid to raise the birds using company-supplied chicks and feed under Perdue's tightly controlled conditions. The fee this farmer earns is calculated through a competition with his peers—the farmers who deliver the heaviest and healthiest birds from a given amount of feed are rewarded with higher pay.

In other words, historically the incentive to use antibiotics was built in. We get out of Stewart-Brown's truck and he leads me to a silo that stores the food Perdue has delivered for this flock. He grabs a laminated "feed ticket" that lists exactly what's inside: protein, fiber, fat, and one additive, narasin.

Narasin is in a class of antibiotics called ionophores, which aren't used in human medicine and are the only antibiotics remaining in Perdue's feed, Stewart-Brown says. "Twelve years ago if you picked this up, it would have said narasin and one or two others," he adds—drugs that might have also been prescribed to people.

About a third of the company's flocks still get ionophores because of a specter that haunts chicken production everywhere: coccidiosis, a parasitic intestinal condition that can trigger severe digestive trouble and cause chickens to stop growing. The industry has relied on ionophores to control coccidiosis for decades. While most researchers I've spoken to don't see ionophores as a major health threat, there is evidence bacteria that evolve to withstand them can also resist an antibiotic used by people, bacitracin. So Perdue is phasing out ionophores in favor of a vaccine for coccidiosis.

Other than the remaining ionophores, Perdue only uses antibiotics to treat a sick flock—typically about 4 percent of its birds each year. As of early 2016, two-thirds of the 676 million birds Perdue slaughters every year never get a drop of antibiotics in their entire lives, not even narasin.

Stepping into the 20,000-square-foot barn with its 40,000 chickens, I brace for an unbearable stench. Instead I get a mild barnyard manure smell and a much stronger, toasty, sweet aroma of chicken feed—a mix of corn and soybeans that I remember from my own days tending a small outdoor flock. There is no stench, I realize, because the chicken house is well ventilated and clean, with plenty of straw on the floor to absorb manure. The room is divided lengthwise by several long, slender feeders full of the corn-and-soybean mix, and by low pipes from which hang metal water nipples every few inches. I move into the massive structure and thousands of birds serenade me with a steady din of clucks; as I walk by, they part as if I'm Moses in a feathery sea.

There amid an ankle-high vortex of fowl, I wonder: If these chickens aren't eating antibiotics to spur weight gain, how do they get plump enough to ensure Perdue's profits and some money for its farmers?

Stewart-Brown surprises me with his answers. The first is the opposite of antibiotics: probiotics, or live cultures (think of the *acidophilus* in yogurt) that can increase the good microbes in the gut, crowding out the bad ones. Probiotics boost the chickens' immunity and even their growth rates, Stewart-Brown says. "Industry guys like to make fun of probiotics—I was one of them, five or six years ago," he says. Back then, he adds, some probiotics marketed for use in chickens were "foo-foo dust," but Perdue found that some actually work. For proprietary reasons he won't tell me which ones, but he explains that they seem to work by shifting the composition of the microbiome—the trillions of microorganisms that live in the guts of chickens and humans alike. (Emerging research suggests the microbiome influences everything from our immune systems to how we metabolize food, which is why antibiotics might be linked to the obesity epidemic.)

After Perdue bought an organic-chicken company called Coleman Natural Foods in 2011, it adopted another unorthodox therapy: oregano. The fragrant herb not only goes well *with* chicken; it also has antimicrobial properties that, when added to feed, help the birds stave off infections. But, I ask Stewart-Brown, won't bad microbes develop resistance to oregano too? Likely yes, he says, so Perdue only uses oregano to prevent particular infections, not as a constant additive.

Moving away from antibiotics, Stewart-Brown says, has forced him to think about the birds' overall well-being—not traditionally a key concern in an industry that profits by converting feed into meat as cheaply as possible. Buying Coleman, he says, "woke up the bird side of us. We'd sort of shut that part down." Now his staff asks questions that sound ripped from *Portlandia:* "Do the birds get what they want?" "Are they healthy?"

Perdue even turns off the lights in the chicken houses for four hours a night so the birds can rest. In the past, lights were left on 24 hours per day on the theory that chickens kept awake eat more and thus get fatter faster. Reducing stress by letting the birds rest, Stewart-Brown says, makes them healthier—and since healthy birds grow faster, the extra sleep has the same effect as constant feeding.

At this point I can't help wondering why Perdue would go way beyond federal recommendations and leapfrog its competitors by switching from dangerous feed additives to finding its "bird side" with oregano and gentler lighting schemes.

Turns out I'm not the only one asking. Dr. Bob Lawrence, the director of the Center for a Livable Future at Johns Hopkins, which has generated reams of research on the dangers of routine antibiotic use on farms, ran into CEO Jim Perdue at a conference recently. Lawrence asked Perdue what had driven the company's flight from antibiotics. "I was hoping he would say, 'The research coming out of your center,'" Lawrence says. Instead Perdue credited worried consumers.

Stewart-Brown says he and his colleagues saw an increase in queries about antibiotics from consumers starting in 2002—perhaps not coincidentally a year after the publication of *Fast Food Nation*, Eric Schlosser's bestseller about the dark side of the U.S. food system. Far more than its competitors, Perdue has a history of courting consumers directly. For decades, while rivals sold essentially unbranded chicken to supermarkets, Frank Perdue peppered consumers with quirky TV ads featuring himself—a balding, skinny pitchman who repeatedly set the quality of his chicken apart from his competitors'. His tagline: "It takes a tough man to make a tender chicken." The ads gave the company a human face, Stewart-Brown says, making people feel invested. Perdue learned to take its customers' concerns seriously: "You can drown them with science to suggest they shouldn't be worried, but the worry is real."

He adds, though, that Perdue executives "could feel" federal regulatory changes coming and wanted to be prepared. In 1996 the FDA had started monitoring levels of antibiotic-resistant bacteria on retail meat, finding it in abundance. Five years later, right around when consumers started complaining, the agency came out with a vague set of recommendations for poultry veterinarians that called for "judicious" use of antibiotics. Those recommendations suggested a regulatory reckoning might be in the offing.

In 2002, Stewart-Brown published a paper in a small academic journal describing a massive experiment that he and two Perdue colleagues had started in 1998. Their research involved nearly 7 million birds raised over three years in 19 farms in North Carolina and Delmarva. The team took away growth-promoting antibiotics from half the chicken houses in both locations and then compared those birds' weight gain and lifespans with those of chickens kept on antibiotic-laced feed.

The differences in weight gain between the two sets of birds were tiny: in Delmarva the antibiotic-free birds weighed on average just 0.03 pounds lighter than their peers; in North Carolina they were just 0.04 pounds lighter—not much for chickens that end up weighing in at about five to six pounds.

I ask Stewart-Brown if the study was the impetus for Perdue going antibiotic-free systemwide. He says the results were key in demonstrating that you could get rid of the drugs without destroying profitability. When consumer demand for antibiotic-free birds started ratcheting up around the same time, he says, the company felt "confident" it could make the shift.

That confidence looks justified now, as accumulating evidence shows the growth-promoting power of antibiotics has been declining for decades. A 2015 Organisation for Economic Co-operation and Development review found that antibiotics sped up growth by up to 15 percent before the 1980s but only by about 1 percent after 2000—a change likely due to better nutrition, hygiene, and breeding. Perdue may simply have calculated that the benefits of using antibiotics for growth were too small, especially in the face of consumer demand and potential regulatory pressure.

Christopher Leonard, a former Associated Press agribusiness reporter who has written a book about the poultry industry, says Perdue was looking for ways to differentiate itself from bigger rivals: "They're just trying to capture market share." Joe Sanderson, the CEO of one of Perdue's rivals, Mississippi-based Sanderson Farms, agrees. His company slaughters 7 percent of the chicken eaten in the United States, making it about the same size as Perdue. "Frankly, these people are doing it for marketing purposes," he tells me. Sanderson, by contrast, has held to the old-school party line, maintaining that "there is no evidence that using these antibiotics for chickens leads to resistant bacteria." Cost is the number-one decision-maker when people go to the grocery store to buy chicken, he says, and using antibiotics remains the cheapest way to produce a lot of meat fast. "We believe the majority of chicken sold in grocery stores will continue to be grown with antibiotics," he says.

Perdue sees a very different future. After 14 years, Stewart-Brown tells me that the company can now produce meat without antibiotics as fast and efficiently as it once did with them. Today Perdue ships two kinds of chicken. About two-thirds of the company's product is already labeled "no antibiotics ever." It sells for about 20 percent more than the remaining third, still treated with antibiotics. Overall, Stewart-Brown explains, Perdue spends an extra $3 to $4 for every $1 it saves in antibiotics reduction—but recoups those costs by charging shoppers a premium for that meat. Last year the company's sales grew faster than those of its competitors—even when the overall market declined.

What's more, Perdue got ahead of the curve, according to Brett Hundley, a vice president and research analyst at BB&T Capital Markets. In recent years, as U.S. sales at fast-food restaurants have stagnated, Chick-fil-A (2014), Subway (2015), and McDonald's (2015) have all vowed to stop serving chicken treated with antibiotics that might also be used for humans. Last year California passed a law that bans the use of antibiotics to promote growth and dramatically tightens the loophole for "prevention" of disease that was left open by the FDA's 2011 guidelines. (Across the Atlantic, a European Union ban on growth-promoting antibiotics took effect in 2006.)

The rest of the chicken-producing giants (save for Sanderson) are scrambling to catch up. Not long after McDonald's announced it would go antibiotic-free, Tyson, the nation's largest chicken supplier, declared it would rid its flocks of all human-important antibiotics by September 2017. Sasha Stashwick, a policy analyst at the Natural Resources Defense Council, says this announcement marked a "tipping point for getting the chicken industry off antibiotics."

Will the changes to Big Chicken ripple through the rest of the meat industry? The FDA doesn't break down data about antibiotic use by species, but given that it grew 23 percent between 2009 and 2014, even as poultry producers began to dial back, it seems pork and beef producers were ratcheting up.

Hundley says antibiotic-free beef and pork command a hefty premium of between 30 and 50 percent. Even so, few producers are taking advantage of that incentive. To go fully antibiotic-free would take time because, he explains, a chicken's short life cycle— about a month from when they are born to slaughter—allows the farmers to quickly see what is working and what isn't. "There's just much more control with chicken," he says. Pigs take six months from birth to slaughter, and beef cows take a year and a half, so making changes takes longer and is riskier. Hundley estimates it will be 15 years before half of all U.S. pork is antibiotic-free and 20-plus years for beef.

And yet there's starting to be some movement. In February, citing concerns about resistance, the food giant Cargill began to reduce the use of medically important antibiotics by 20 percent in about one-fifth of its cattle. Or take Smithfield Foods, by far the biggest U.S. pork producer, now owned by the Chinese conglomerate WH Group. On its website the company says it still uses medically important antibiotics "to control, treat, and prevent disease," but not "to promote growth or for feed efficiency." Admittedly, it's hard to monitor why producers dose their livestock with drugs. A 2014 analysis by the Pew Charitable Trusts found that 66 of the 274 medically important antibiotics that the FDA allows to be used for disease prevention can also be useful in growth promotion.

The truth is, antibiotics have been an easy fix for an industry under enormous pressure to produce maximum amounts of animal protein at minimal cost. Levy's studies showed the dangers of doing that nearly 40 years ago, yet until the recent exception of Perdue, neither regulators nor leading producers chose to act.

Meanwhile, even as global trade has made us more susceptible to superbugs that spring up half a world away, we've exported our antibiotic addiction to countries unlikely to exercise caution.

Over the past 15 years, China has been rapidly scaling up its meat production, and farmers there have embraced drugs with little oversight. A 2013 analysis by a Beijing-based agribusiness consulting company found that more than half of all antibiotics in China were used on livestock. That's a smaller ratio than you'll find in the United States, but China is expected to double the amount of drugs it feeds to animals by 2030. Last year Chinese researchers shocked public health authorities when they announced in *The Lancet* that they had found a strain of *E. coli* in pigs and humans that had evolved to withstand colistin, a potent antibiotic widely considered a last resort against many multidrug-resistant pathogens. Worse, the gene that allowed the *E. coli* to shrug off colistin easily jumps to other bacterial species and is "likely to spread rapidly into key human pathogens" like salmonella and Klebsiella —a bacteria that can cause infections like pneumonia and meningitis. The authors warned that these colistin-defying pathogens were "likely" to go global.

And in a particularly perverse disincentive, large pharmaceutical companies are no longer keen to invest in new antibiotics. The reason is that, according to a World Health Organization report, resistance sets in too fast for them to make money off new drugs. So even if consumer pressure forces the rest of Big Ag to follow Perdue's lead, the industry's 40-year delay may still end up costing countless lives.

Bob Lawrence of Johns Hopkins says a more optimistic future relies on two conditions: public investment in developing new antibiotics and a binding global pact to severely ramp down farm antibiotic use. Unless we start rolling out new drugs and using the old ones sparingly, he says, "the genie is so far out of the bottle that we're facing a rather bleak future."

NATHANIEL RICH

The Invisible Catastrophe

FROM *The New York Times Magazine*

"IT JUST SEEMS like a beautiful day in Southern California,"
Bryan Caforio said.

It was late January in Porter Ranch, an affluent neighborhood
on the northern fringe of Los Angeles. Caforio and I sat at a Starbucks overlooking an oceanic parking lot crowded with shoppers.
The air was still, dry, 70 degrees. Caforio, a young trial lawyer
running for Congress in the state's 25th District, gestured at the
pink and orange striations of sky above Aliso Canyon, its foothills
bronze in the falling daylight. "It seems like a beautiful sunset in
a wonderful community," Caforio said, "and we're sitting outside,
enjoying a wonderful coffee."

But there were scattered clues that suggested that everything
was not so wonderful. Near a trio of news vans parked in front of
the Starbucks, antenna masts projecting from their roofs, a camera-
man stared quizzically up at the canyon. Next to the SuperCuts, security guards stood outside two nondescript storefronts; stenciled
on the windows were the words COMMUNITY RESOURCE CENTER
and, in smaller letters, SOCALGAS. The guards asked for identifica-
tion and dismissed anyone who tried to take a photograph. At the
entrance to Bath & Body Works, a device that resembled an electronic parking meter was balanced on a tripod; the digital display
read BENZENE, followed by a series of indecipherable ideograms.
The parking lot held a preponderance of silver Honda Civics bear-
ing the decal of the South Coast Air Quality Management District.
Inside the cars men sat in silence, waiting.

Beyond the Ralphs grocery store and the Walmart rose a

neighborhood of jumbo beige homes with orange clay-tiled roofs and three-car garages. The lawns were tidily landscaped with hedges of lavender, succulents, cactuses, and kumquat trees. The neighborhood was a model of early-1980s California suburban design; until October it was best known for being the location where Steven Spielberg shot *E.T.* But now the meandering streets were desolate, apart from the occasional unmarked white van. As you ascended the canyon, reaching gated communities with names like Renaissance, Promenade, and Highlands, the police presence increased. On Sesnon Boulevard, the neighborhood's northern boundary, an electric billboard propped in the middle lane blinked messages: REPORT CRIME ACTIVITY; LAPD IN THE AREA; CALL 911. Holleigh Bernson Memorial Park was empty aside from three cop cars, patrol lights flashing.

But the most significant clues were the spindly metal structures spaced along the ridge of the canyon. They resembled antennas or construction sites or alien glyphs. Until recently most residents of Porter Ranch did not pay them much attention.

"You look at the hills, you see a few towers," Caforio said. "But do you really know what they are?" He shook his head. "You try to say, Hey, we're having an environmental disaster right now! But it just looks like a beautiful sunset."

The first sign of trouble came on October 25, when the Southern California Gas Company filed a terse report with the California Public Utilities Commission noting that a leak had been detected on October 23 at a well in its Aliso Canyon storage facility. Under "Summary" the report read: "No ignition, no injury. No media."

The local news media began to take notice, however, when Porter Ranch residents complained of suffocating gas fumes. In response, SoCalGas released a statement on October 28 pointing out that the well was "outdoors at an isolated area of our mountain facility over a mile away from and more than 1,200 feet higher than homes or public areas." It assured the public that the leak did not present a threat.

Timothy O'Connor, the director of the Environmental Defense Fund's California Oil and Gas Program, had read about the complaints. But he did not think much of them until November 3, when, at a climate-policy event in downtown Los Angeles, he learned from an acquaintance who worked at SoCalGas that the company was flying in experts from around the country to help plug the leak. At home that night, O'Connor read everything he could about the Aliso Canyon gas field. How much gas was stored inside the canyon? How much could leak out?

The foothills on which Porter Ranch was built, O'Connor learned, once belonged to J. Paul Getty. His Tide Water Associated Oil Company hit crude in 1938 and did not sell the land until the early 1970s, after it had extracted the last drop. The drained oil field was bought by Pacific Lighting, which used it to store natural gas. With a capacity of 84 billion standard cubic feet, the cavity, which lies between 7,100 and 9,400 feet below the surface, is one of the country's largest reservoirs of natural gas (which is composed mainly of methane). The facility functions as a kind of gas treasury. When prices are low, the company hoards the gas inside the canyon; when they are high, it releases the gas into pipelines that snake through Los Angeles, heating homes, fueling stoves, and providing power to solar- and wind-energy facilities.

The 115 wells in Aliso Canyon can be imagined as long straws dipping into a vast subterranean sea of methane. The leaking well, SS-25, is a steel tube seven inches in diameter that descends 8,748 feet from the canyon's ridge. The well is plainly visible from many of the streets in Porter Ranch. From the ground it resembles a derrick set beside a series of low white buildings. If you look at it through a pair of binoculars, you can make out, flying from its highest girder, an American flag.

After conducting some basic calculations, O'Connor arrived at a shocking conclusion. Given the pressure and quantity of gas stored within, the canyon was like an overinflated balloon; a puncture could release in a single day as much gas as 1,785 houses would consume in a year. As it turned out, O'Connor was mistaken —the figure ended up being much higher than that—but he included it in an urgent letter he sent the next day, on November 4, to the governor's senior energy adviser and members of the California Air Resources Board, Public Utilities Commission, Energy Commission, and Department of Conservation. He demanded that the agencies conduct "an accurate and public accounting of the gas lost at Aliso Canyon."

That evening O'Connor attended a hearing at the Community School in Porter Ranch with about 100 panicked residents. They complained that the gas fumes were causing headaches,

respiratory problems, nosebleeds, and vomiting. The next morning, having yet to receive a response to his letter, O'Connor realized that he didn't have to wait for the state to take action. He
could call Stephen Conley.

Conley is an atmospheric scientist at the University of California, Davis and the founder of Scientific Aviation. He flies a single-
engine Mooney TLS that looks like something Cary Grant might
have flown in *Only Angels Have Wings*. Public agencies, scientists,
and nonprofit organizations that study the climate hire Conley to
loop over oil and gas fields at low altitudes, measuring methane
concentrations with a device called a Picarro analyzer. At 10:30
a.m. on November 5, Conley took off from Lincoln Regional Airport, just north of Sacramento. As a courtesy, O'Connor notified
Jill Tracy, the director of environmental services at SoCalGas. He
began to receive a flurry of text messages from executives at SoCal-
Gas. They said the flight was unsafe and inappropriate. But SoCal-
Gas was not concerned for the safety of the pilot, as O'Connor first
assumed. The executives claimed to be concerned for the workers on the ground, who were operating cranes and drills in an
effort to plug the leak. The workers, Tracy wrote, might become
distracted by the sight of an airplane overhead, with catastrophic
consequences.

O'Connor found this reasoning odd, because Porter Ranch lies
in the flight path of Van Nuys Airport. Nearly 600 flights take off
or land there every day. He proposed that the airplane keep one
mile away from the well site. SoCalGas executives said they still
considered this unsafe. O'Connor asked whether there was a safe
distance from the well at which the airplane could fly. The company said there was not. Conley was forced to turn back.

Two days later, though, Conley was back in the air, this time
on assignment for the California Energy Commission. Over the
course of the next four months Conley flew 15 flights over the site.
On his first flight Conley's Picarro analyzer registered 50 parts per
million. The normal concentration of methane in the atmosphere
is two parts per million. Conley thought something was wrong with
the instrument. But a backup analyzer gave the same reading. He
recalled, "That's when I said, Oh, my God, this is real."

What is real to a climate scientist is abstract to the rest of us. The
study of the climate is a study of invisible gases. In order to trans-

late findings to a public lacking a basic understanding of atmospheric chemistry, climatologists must resort to metaphor and allegory. They must become writers, publicists, politicians. This doesn't always come easily. The leak at Aliso Canyon, Conley discovered, was the largest methane leak in the country's history. But what did that mean?

You could begin by comparing emissions from the gas leak at Aliso Canyon with other pollution sites. Conley had logged about 1,500 hours of flight time over oil and gas fields, moonscapes like the Barnett and Eagle Ford Shales in Texas, the Julesburg Basin in Colorado, and the Bakken Formation in North Dakota. The highest methane-emission rate he had ever recorded was three metric tons per hour. The methane was leaking from Aliso Canyon at a rate of 44 metric tons per hour. By Thanksgiving it had increased to 58 metric tons per hour. That is double the rate of methane emissions in the entire Los Angeles Basin. This fact takes some effort to absorb. It means that the steel straw seven inches in diameter plugged into Aliso Canyon was by itself producing twice the emissions of every power plant, oil and gas facility, airport, smokestack, and tailpipe in all of greater Los Angeles combined.

In a paper published in the February issue of *Science,* Conley and his coauthors estimate that 97,100 metric tons of methane escaped the Aliso Canyon well in total. Over a 20-year period, methane is estimated to have a warming effect on Earth's atmosphere 84 times that of carbon dioxide. By that metric, the Aliso Canyon leak produced the same amount of global warming as 1,735,404 cars in a full year. During the four months the leak lasted—25 days longer than the BP oil spill in the Gulf of Mexico—the leak contributed roughly the same amount of warming as the greenhouse-gas emissions produced by the entire country of Lebanon. If well SS-25 were a nation-state, it would have contributed to global climate change at a rate exceeding that of Senegal, Laos, Lithuania, Estonia, Zimbabwe, Albania, Brunei, Slovenia, Nicaragua, Panama, Jamaica, Latvia, Georgia, Guinea, Equatorial Guinea, Costa Rica, Honduras, Tajikistan, Armenia, and Iceland. SS-25 would fall just behind Mali.

These facts, despite their world historical significance, still failed to make much of an impression locally and nationally, let alone internationally. What was one more airborne toxic event at a time when the global climate was itself an airborne toxic event? The

World Health Organization has called climate change the greatest
global health threat of the 21st century, an opinion shared by the
United Nations, the Environmental Protection Agency, and the
National Institutes for Health, among others. By 2030 increased
rates of heat stress, infectious-disease transmission, and malnutrition caused by climate change are expected to cause an additional 250,000 deaths a year. Yet as gargantuan as the Aliso Canyon
emissions might be, their influence on the climate would have no
immediate or direct effect on the lives of the residents of Porter
Ranch. Residents were as concerned about the leak's contribution
to atmospheric warming in the years and centuries to come as everyone else on the planet—which is to say, not especially. We are
already immersed in leaking invisible gases with largely invisible
effects too overwhelming to control. What difference was another
Lebanon's worth of emissions?

The residents of Porter Ranch were very concerned, however,
about what the inhalation of the gas might do to their brains
and their lungs. Some residents found the smell of gas so overwhelming that they sealed their windows and doors and refused
to go outside. Others could not smell the gas and experienced
no symptoms. Sometimes those with severe symptoms and those
without lived in the same household. In the absence of reliable information from SoCalGas or state agencies, the residents of Porter
Ranch underwent their own transformation: they became amateur
scientists, epidemiologists, sociologists, political theorists. They began to develop their own hypotheses.
    "Yellow spots," Charles Chow said, "are coming out of the atmo-
sphere."
    I met Chow, a 76-year-old retiree with mirthful eyes and springy
joints, in his driveway in late January. He was installing new shocks
on his 1992 burgundy Cadillac Brougham Elegante. Beside the
Cadillac was a 1986 Silver Spirit Rolls-Royce. In the street, which is
called Thunderbird Avenue, there was a 2002 Black Thunderbird.
Chow pointed out the spots. They were about the size and color
of a yellow split pea. They had appeared on the windshields of his
cars, on the Cadillac's vinyl roof, on the canyon-facing windows of
his home.
    Chow first became concerned about the Aliso Canyon leak in

October, when Chaka Khan, his Chihuahua/miniature pinscher, began having severe respiratory problems. His wife, Liz, began suffering chronic headaches, eye irritation, and a sore throat. Her doctor said there was nothing he could prescribe her. The only thing she could do, he said, was to leave Porter Ranch. Most of their neighbors fled before Thanksgiving. On their block alone, Chow estimates that 15 households, mostly retirees, relocated. Since then the Chows have driven four times a month to a vacation rental they share on the Baja Peninsula, 60 miles south of the border, "just to get out of the atmosphere," Chow said. In the Mexican air, Liz's symptoms vanished.

Chow was soon joined in his driveway by Rick Goode, a neighbor of 25 years with a slender build and a birdlike gait. Goode wanted Chow's advice about legal representation: about two dozen plaintiffs' firms had descended on Porter Ranch since October, competing to sign as many clients as possible. What did Chow think of Robert Kennedy's firm? Or Weitz & Luxenberg, which had sent Erin Brockovich to solicit clients? The previous week Brockovich told reporters that she "started feeling kind of dizzy" within 10 minutes of arriving in Porter Ranch. Chow ruled her out.

"You don't get sick that fast," he said.

"I've been having terrible headaches," Goode said. "Have you?"

"My wife has headaches every day, sore throats," Chow said. "I don't. We both live in the same house. Everybody is different."

Liz returned from a doctor's appointment. She removed her sunglasses to reveal a new cyst on her eyelid. She searched for a word to describe her general condition since October. "A malaise," she said finally.

Barbara Weiler, 64, who was walking her dog very slowly several blocks away, first experienced the malaise in gym class. "You felt like you were lazy," she said. "It was obvious when we were using the resistance bands. We felt like we didn't want to work as much as we normally would."

Paula Vasquez found the smell of gas so strong in late October that she was certain there was a leak inside her house. She hasn't opened a window since. She and her family—she lives with her husband, their 33-year-old daughter, and their 13-year-old grandson—have experienced bloody noses, blurred vision, and nasal congestion. But Vasquez has also noticed other signs. She pointed to fruit trees in her neighbors' backyard. "I see them picking lemons," she said. "I don't say anything, but I'm concerned for them. Is there gas in the fruit?"

She showed me photographs she made her grandson take on her cell phone while she was driving home on the Ronald Reagan Freeway. In the sky above Porter Ranch, a heavy funnel of clouds was lit neon orange.

"It looks like a big atomic cloud," she said. Vasquez had a warm, cheerful manner; horror did not come naturally to her. "Creepy, huh? But I don't know anything about science."

We are a show-me species, wired to look for visible evidence of invisible harm. That impulse can lead a person to blame global warming for a hot day in February or, conversely, make a climate-change denialist find vindication in a snowstorm. But the world's largest natural-gas leak has no known effects on clouds or lemons. (It may, however, create yellow dots. Michael Jerrett, the director of the Center for Occupational and Environmental Health at UCLA, explained that the dots are most likely a residue of the petroleum-laced slurry used to plug the leak.) The most dangerous threats to our species are precisely those that are most difficult to visualize: long-term, slow to emerge, amorphous. These threats include not only warming temperatures but also mutating viruses and political corruption and tend to be invisible, dimensionless, and pervasive, like death. Like natural gas.

While the yellow dots were coming out of the atmosphere and staining the vinyl roof of Charles Chow's Brougham Elegante, the planet was enduring the warmest January on record. It was the fourth consecutive month in which global temperatures beat historical averages by more than one degree Celsius, another novelty. This news, when publicized at all, tended to be accompanied by NASA's map of the world, overlaid muddily with orange and red splotches denoting temperature increases; otherwise there appeared stock photographs of sunbathers on beaches or icicles melting. Then came February, the warmest month in recorded history. The threat to human civilization is advancing faster than ever before—the climate is warming faster than at any time in the last 65 million years—but all we can see are sunbathers and melting icicles.

All that the residents in Porter Ranch could see during those months of yawning uncertainty were empty streets and mysterious white vans. They were desperate for answers: Was the gas making them sick? How could they protect themselves? Who would be held responsible? The personal-injury lawyers were well prepared. They offered clarity, assurance, optimism. They could predict, with confidence, the future—a profitable future for the residents of Porter Ranch. Since November the firms had been holding weekly informational meetings at local churches and hotels. At each session lawyers answered questions from the community, often for several hours at a time, and circulated client forms.

Rick Goode and the Chows attended one such meeting in late January, two days after their driveway conversation on Thunderbird Avenue. It was hosted by R. Rex Parris at the Hilton in Woodland Hills, about 10 miles south of Porter Ranch. R. Rex Parris belongs to a consortium of law firms that on December 10 filed the first class-action complaint against SoCalGas and its parent company, Sempra Energy, the nation's largest natural-gas utility, on behalf of hundreds of homeowners. The group's news release anticipated that the leak would end up costing Sempra shareholders "well over $1 billion." On this morning, about 20 Porter Ranch residents sat at conference tables, grazing on the free coffee, doughnuts, and bagels. A young lawyer, who seemed to have consumed a large quantity of the coffee, stood at the front of the room, delivering her sales pitch.

"Anything SoCal tells you," she said, "don't listen to it. Everything they say means nothing."

She advised the residents to keep daily journals. They were to note each occurrence of a physical symptom or a gas smell and list all expenses incurred by relocation or illness. Someone asked whether he could qualify as a plaintiff even if he lived 10 miles from Aliso Canyon.

"Nothing's been established yet," the lawyer said. "I've heard between five and 10 miles. But we don't have the data yet."

"They claim it started on October 23," one older woman said. "But in April my dog, a boxer, died within two weeks. I know it was the gas."

"It was earlier than October 23," the lawyer said. "I just don't know when. We want it to be as early as possible, so we can get much more money for everyone."

The residents nodded in approval.

The lawyer explained that about 30 attorneys were assigned to the case. The firms would receive as their fee 30 percent of any payouts. "We're predicting a settlement," she said.

A Russian man who resembled Gérard Depardieu exclaimed, "I escaped Chernobyl for this!"

The man, Igor Volochkov, later told me that in 1986 he moved from Kiev, 60 miles south of Chernobyl, to Los Angeles when his wife was pregnant with their son. "We ran away to save our lives and the lives of our children."

Volochkov said he knew something was wrong in October, when his parrot, Bon, dropped dead. He bought a new parrot and a parakeet, Gosha and Margosha, but they died within a month. The same thing happened in Kiev, when, he said, nuclear radiation from Chernobyl killed his parakeet Petruschka. Volochkov said his son had asked why they moved from one Chernobyl to another Chernobyl.

"Maybe," Volochkov replied, "our destiny is to fight against Chernobyls."

In mid-November, nearly a month into the leak, the Los Angeles County Department of Public Health ordered SoCalGas to pay for new housing for anyone affected by the gas odors. Nearly 6,000 households, about half of Porter Ranch's population, accepted the offer, moving to hotels, apartments, and houses in surrounding neighborhoods. Those who did not relocate immediately struggled to find available short-term rental properties, but others took advantage of the gas company's largess. Charles Chow said he knew a family paying $10,000 in monthly rent. "People are gouging the gas company," he said. "I don't believe in unfair practice. I was a businessman. I think fair is fair. All I might take is reimbursement for money I've had to pay the vet for Chaka Khan."

Jerry McCormack, another neighbor of the Chows, has rarely detected an odor and has not been sick. "I think there's a lot of foolishness going on," he said. "This is not Fukushima. The rental market has gone crazy. Everyone is out to get the gas company. The hysteria is proportional to the number of lawyers coming to town." He conceded that his wife, who is recovering from cancer, "can smell it quite well" and is concerned. Her oncologist advised her to leave.

Adam and Mindi Grant, a couple in their mid-40s, live a mile from the leak site. Their three children play basketball and swim outside. "We've legitimately smelled it one day," Adam said. "We joke about it. Every time someone gets a bloody nose, we say, It's the gas!"

Adam teaches world history at a local high school. Mindi is an insurance lawyer. "I have friends with real symptoms," she said. "Some, maybe not. They're setting up for a money grab. They think there's big money, deep pockets. But they're going to have trouble showing causation."

"Had the smell been horrific," Adam said, "we would have relocated. But because it's not affecting us as a family, I'm a little lackadaisical about it."

If the smell of gas makes one person dizzy while the neighbor next door can't smell anything, is one of them lying? If a man does not actually inhale gas but develops headaches and nausea anyway, is his suffering any less? Disaster psychiatrists call this phenomenon *somatization,* a word that has replaced *hysteria* and *psychosomatic,* terms now considered offensive. "In manmade disasters, the psychological consequences can be very severe and ongoing," David Eisenman, the director of the Center for Public Health and Disasters at UCLA, told me. "Unexposed individuals can have symptoms similar to people who have exposure." Fear makes you sick. As it turns out, inhaling poisonous gas causes the same symptoms as the fear of inhaling poisonous gas: headaches, dizziness, and nausea.

Porter Ranch residents had reason to be afraid. Nobody could tell them what they were breathing. Methane was gushing from the leak, of this they could be certain, but methane was not what they smelled. Methane is odorless. What they smelled were mercaptans: sulfur compounds that in nature are released in animal feces. Mercaptans are added to natural-gas pipelines to provide an olfactory alarm in case a leak occurs, the way banks insert exploding dye packs into bags of cash. Inhalation of mercaptans can cause headaches, dizziness, and nausea, but like methane, they are not currently known to cause significant long-term health effects. The main health concern about the leak was that other, more toxic gases might also be escaping from the bowels of Aliso Canyon—including gases remaining from its previous life as an oil field.

Chief among these was benzene, a known carcinogen. Los Angeles air, among the most polluted in the nation, tends to have a background concentration of benzene between 0.1 and 0.5 parts per billion. The World Health Organization has declared that "no safe level of exposure can be recommended." In November readings taken by SoCalGas near its facility found benzene concentrations fluctuating wildly between 0.3 ppb and a nightmarish 30.6; readings taken by the company in Porter Ranch shot as high as 5.5 ppb. Other toxic gases—toluene, xylene, hexane, and hydrogen sulfides—were also detected at higher-than-normal concentrations. The South Coast Air Quality Management District also tested the air quality in the first two months of the leak, but the monitoring was sporadic and conducted at only a handful of locations in the community. By mid-January, after efforts to depressurize the well had managed to reduce the leakage rate considerably, a more rigorous study by Michael Jerrett of UCLA found that the air in Porter Ranch was in fact unusually clean—most likely because of the absence of so many residents and their cars.

California health officials believe that there will be no long-term health effects from the leak. "Increased cancer risk is very small," said Dr. Melanie Marty, the acting deputy director for scientific affairs for California's Office of Environmental Health Hazard Assessment. "Much smaller than routine risks we experience every day." But Jerrett suspects that during the first six weeks of the leak, when the gas escaped at a much higher rate, conditions might have been dangerous, particularly for children and older residents. On March 10, following complaints from relocated residents who suffered nosebleeds and skin rashes after moving back home, Jerrett took dust samples at seven houses in Porter Ranch. Two contained benzene and hexane, a finding that Jerrett found "concerning."

The actual composition of the gas was only the beginning of what the residents of Porter Ranch did not know about the invisible fumes seeping from Aliso Canyon. They did not know how far the gas was drifting or in what quantities. It seemed that the smell was stronger the higher you went up the mountain and stronger at dusk and dawn, but there was little data to support this. There was also the mystery of the complex local wind patterns, which resemble those of no other part of the Los Angeles Basin and change direction capriciously.

No one even knew what had caused the leak in the first place, though a broken safety valve, removed by SoCalGas in 1979 and never replaced, received some blame. In 2012, President Obama signed a pipeline-safety bill that should have prevented a leak of this kind. But in Aliso Canyon the new regulations were not enforced. "We have the law, but no one is complying," said Mel Reiter, the editor of *The Valley Voice,* a monthly newspaper that may be the only local business to profit from the leak: more plaintiffs' law firms sought full-page ads than it has pages. "There are 115 active wells, and more than two-thirds were built before 1980," Reiter said. "If one is leaking, what are the odds that 30 more are, or will soon?"

Regulations are in place, but nobody knows who can enforce them. When Matt Pakucko, a lead plaintiff in the class-action lawsuit, first smelled the leak on October 24, he called SoCalGas. He was told that the company was merely "releasing gas into the air," which was "something that they do periodically," and that there wasn't a leak. He knew that the South Coast Air Quality Management District was responsible for investigating air-quality complaints. But SoCalGas, a private utility, did not fall under the regulatory oversight of any single agency. Besides the Air Quality Management District, agencies responsible for responding to the leak included the State Energy Commission, the Los Angeles County Department of Public Health, the Air Resources Board, the Public Utilities Commission, the Division of Occupational Safety and Health, the Department of Conservation's Division of Oil, Gas and Geothermal Resources, the Environmental Protection Agency, the Office of Environmental Health Hazard Assessment, the County Fire Department, and the Governor's Office of Emergency Services. In January the Los Angeles County Board of Supervisors called for the creation of yet another regulatory "structure" to oversee gas-storage facilities.

For most Porter Ranch residents, all this confusion added up to a single fact: an invisible gas was threatening their lives. "We don't know what methane is," said Sam Kustanovich, a Belarussian pawnbroker who had the misfortune of buying his house two months before the leak was detected. "Nobody knows. It could mean explosions. Me, I'm afraid of explosions."

The global climate, even in drought-stricken Southern California, is not an especially consequential campaign issue. A menacing disaster that causes mass vomiting and mass nosebleeds in a wealthy, vote-rich community, however, is a candidate's dream. In this election season the procession of scientists and lawyers heading to Porter Ranch has been trailed by a caravan of Californian politicians. None have come out in favor of mass nosebleeds. Though the 25th Congressional District reaches only its pinkie toe into Porter Ranch, Bryan Caforio, a Democrat, has made the leak a central issue of the election, which promises to be one of few closely contested races in the House. The Republican incumbent, Steve Knight, who has received campaign donations from Sempra Energy, said in December that he was confident that SoCalGas was "working on this as diligently as they can" but more recently called for a congressional hearing on the matter and introduced safety regulations for natural-gas storage. Even the Los Angeles County supervisor, Michael Antonovich, a Republican who has voted consistently against regulation efforts, has loudly proclaimed his determination to hold SoCalGas responsible.

"We're all kind of feeding on it in a weird way," said Henry Stern, a Democrat who is running for state senate in the local district. He previously served as an adviser on energy and climate policy for the district's current senator, Fran Pavley, a Democrat who cannot run again because of term limits. "How often are there climate disasters in suburbia?"

Stern has been struck at community meetings by the comments of local residents, many of them self-identifying conservatives, who have begun to question the wisdom of relying on fossil fuels. "Climate change is not a real thing for most of these people," Stern said. "But you change your mind quick when your kids are puking."

The only politician who has failed to use the gas leak for political gain is Governor Jerry Brown. His Office of Emergency Services, following protocol, began monitoring the leak in October and began coordinating the state's response in mid-November, overseeing the various state agencies responsible for responding to it. On December 18, Brown sent a stern letter to the chief executive of SoCalGas, urging cooperation and demanding accountability. "Everything that could be done under the authority of the governor was being done," Mark Ghilarducci, the director of the Governor's Office of Emergency Services, told me. But Brown made no public statement until January, when he toured the SoCalGas facility and met privately with four members of the local Neighborhood Council; this was 10 weeks into the leak and a month after he attended the United Nations climate talks in Paris, where he boasted of California's emissions-reduction plan, the most ambitious in North America. Brown declared a state of emergency in Aliso Canyon on January 6, but for many in Porter Ranch, that wasn't nearly soon enough.

"We're suffering because Jerry Brown is so not involved in this," Matt Pakucko said. "There he was in Paris, saying look how green California is, while 10 years of green stuff is going into the air right now."

Ghilarducci disputes this. "This concept that nothing happened and the governor was not engaged until he issued a state of emergency on January 6 is just absolutely not correct," he said. "Let's face it: we deal with so many emergencies out here. This is not Vermont, this is not Oklahoma, where you have a state the size of Sacramento County. This is a nation-state." He continued, "The governor is very confident that he doesn't need to be on the scene, holding a press conference, to show that he's doing something."

The governor's reputation in Porter Ranch was not helped by the revelation that his younger sister, Kathleen Brown, is a paid board member of SoCalGas's parent company, Sempra Energy. "I'm sure there's a conflict of interest," Rick Goode said. "My feeling is it's an 'I scratch your back, and you scratch mine.' It concerns me." In 2013 and 2014, Kathleen Brown received $456,245 in compensation, including stock awards. A partner at the firm of Manatt, Phelps & Phillips, she also has, according to the Public Accountability Initiative, a $949,653 stake in the Forestar Group, a real estate and natural-resources company, where she is a director and major shareholder. Forestar is developing Hidden Creeks Estates, a gated community of 188 luxury homes, right next to Porter Ranch, on property abutting Sempra's.

Kathleen Brown's office at Manatt referred me to Doug Kline, the director of corporate communications for Sempra Energy, who would not give a specific comment on Brown's role. But he said, "Our entire board of directors has been actively engaged and regularly briefed on the Aliso Canyon incident." Deborah Hoffman, Jerry Brown's deputy press secretary, wrote in a statement that any implication that the state did not exercise "its full regulatory and oversight authority" was "scurrilous and irresponsible."

*

SoCalGas announced on February 18 that the well had been sealed. Chris Gilbride, a spokesman for SoCalGas, wrote in an email, "Throughout the incident, air samples for benzene and other compounds were found to be at or near levels seen in the rest of the county and below levels of concern." He continued, "The Los Angeles County Department of Public Health has consistently reported that no long-term health effects are expected due to the leaking well." In late February many residents, including Rick Goode and Igor Volochkov, said they still smelled gas or still suffered symptoms. "Maybe SS-25 is capped," said Kyoko Hibino, Matt Pakucko's girlfriend. "But I think there is still something seeping up from underground. I think other wells are continuing to leak. The smell is still pretty strong. It is out there still."

It is uncertain whether the residents of Porter Ranch will experience health effects in the long term. It is certain that the atmosphere will experience long-term effects. But the effects will be as indecipherable as a plume of colorless gas leaked into a windswept canyon. How do we make sense of the addition to the atmosphere of thousands of tons of invisible gases that will have semi-invisible effects on us and only slightly more visible effects on generations we won't live to see?

"If you compare the Aliso Canyon leak to other leaks," said Stephen Conley, the aviator-scientist, "it's top dog. It's a monster. It throws off LA's emissions for the year. It's a significant percentage of California's annual carbon budget. But it's about 0.002 percent of the global methane budget. It's not like next year will be warmer because of Aliso Canyon."

This is true. It's not like next year will be warmer because of the car trips that Porter Ranch residents make to their temporary rental homes, or the gas they use to cook dinner, or the energy required to heat their swimming pools. Next year won't be warmer because of the 200,000 airplanes passing through Van Nuys Airport. Next year won't even be warmer, necessarily, because of the roughly 140 billion cubic meters of natural gas that oil companies flare into the atmosphere. But next year will be warmer.

CHRISTOPHER SOLOMON

The Devil Is in the Details

FROM *Outside*

THE AMERICAN WEST is our handsome conundrum—too beautiful to use, too useful to be left alone, as a Colorado journalist once put it. In the past the landscape seemed so enormous that conflicting dreams could find room in its whistling emptiness. Now there's not much left that we haven't touched, and we argue about how to manage what remains—a quarrel over whose dreams should come first.

Nowhere is the argument louder than in the creased country of eastern Utah, a place you know even if you've never been there: stone arch and sunburned canyon, perfect desert sky. The area is home to marquee national parks like Arches and Canyonlands, but much more of it exists as sprawls of federal land that taken together are larger than many eastern states. Some people look at the region's deep slots, peaks, and antelope flats and are inspired to protect them from development. Others hunger for what lies beneath: natural gas, oil, and potash.

If conservation is a tricky project in today's rural West, with a resurgent Sagebrush Rebellion leading to events like the armed takeover of a federal wildlife refuge in Oregon, it's particularly confounding in Utah, where for the better part of a century a war over wilderness has been fought. Recently, during travels through small eastern Utah towns like Moab and Vernal and Blanding, I met more than one Utahan whose pioneer ancestors had arrived by wagon train and who still couldn't bring themselves to utter the word *wilderness,* famously defined in the 1964 Wilderness Act as "an area where the earth and its community of life are untrammeled by man, where man himself is a visitor who does not remain." Instead they called it "the W-word," and they spat it out when they said it. In Utah more than other states, environmentalists and their foes wield just enough power to stymie each other. The toll has been great: enemies have grown gray squatting in the same trenches their fathers dug, and still the land remains unconquered by either side.

In early 2013 a conservative Utah congressman named Rob Bishop sent a letter to more than 20 groups in both camps of the wilderness wars. He said to them, simply, Tell me what you want. He proposed using their responses to frame an ambitious "grand bargain" designed to end the state's wilderness disputes. In Bishop's hoped-for compromise—officially known as the Public Lands Initiative—everybody would get some of what they desired, as long as they were willing to negotiate. Conservationists could potentially protect the largest amount of wilderness acreage in a generation—a Vermont-size array of wildlands that would be free from development forever—including places whose names are talismanic to desert rats: Desolation Canyon, Indian Creek, Labyrinth Canyon. And the conservative counties of eastern Utah would get some assurances that more motorized recreation and future development would be allowed to happen on other public lands, along with the economic benefits those projects promise to deliver.

Late last fall, during a telephone interview from his offices on Capitol Hill, Congressman Bishop told me that he decided to pursue the grand bargain when he realized that the old conflict-based manner of tackling disputes had failed. "I've got to break paradigms and try something new," he said. But why something so ambitious in a place so difficult? Bishop invoked a favorite quotation, one that he and others often attribute, perhaps apocryphally, to Dwight Eisenhower: "Whenever I run into a problem I can't solve, I always make it bigger. I can never solve it by trying to make it smaller, but if I make it big enough I can begin to see the outlines of a solution."

Eventually any such plan will be introduced into Congress as a bill that spells out what's being traded for what. And in theory a grand bargain that has the support of all sides would stand a decent shot of navigating the hyper-partisan halls of Congress to become law.

A few years ago, if you had asked which politician would emerge as the Great Compromiser on wilderness, Bishop wouldn't have been anybody's first choice. Now 64, he's a seven-term congressman from Utah's First Congressional District, which stretches across the top of the state. The federal government manages nearly two-thirds of Utah's land, and Bishop firmly believes Uncle Sam is terrible at the job. He has likened federal ownership to Soviet collectivism, and he argues with a pungent eloquence—he's a former high school debate coach—that the government should get out of the stewardship game and revert the land to local management.

"When you try to control the land from a four-, five-hour flight away, the people always screw up," he told me. He has repeatedly fought to weaken environmental laws and neuter federal agencies like the Bureau of Land Management, the National Park Service, and the Forest Service.

Yet Bishop is no wild-eyed back-bencher. He's chairman of the powerful House Committee on Natural Resources, with sway over issues ranging from energy production to mining, fisheries, and wildlife across one-fifth of the nation's landmass. Last fall he helped kill renewal of the 51-year-old Land and Water Conservation Fund, which raised billions for recreation and state and federal land acquisition, until it could be overhauled. (The fund was reauthorized in mid-December, but for only three years.) The League of Conservation Voters has given Bishop a 3 percent lifetime rating. Still, he can be a powerful ally, as his work on the grand bargain shows. Bishop's effort is a genuine attempt to solve the kinds of long-stewing western land-use conflicts that at their worst devolve into potential violence.

Time is running short for Bishop, however. The Obama administration has given him room to cobble together a deal with conservationists, ranchers, Native Americans, energy companies, and others—the kind of huge grassroots pact that most parties would prefer. But if an agreement isn't reached soon, the president appears poised to step in and do some preservation of his own, in the form of a major new national monument in eastern Utah called the Bears Ears. The clock is ticking.

For some critics, though, the deal struggling to be born is a devil's bargain. And it has exhumed a pressing version of an age-old question: How much can you compromise before you sacrifice the very thing you're trying to save?

*

Last September, nearly three years after Bishop sent out his letter, I traveled to eastern Utah to check on the progress of his idea. Initially Bishop had asked various groups—everyone from the Utah Farm Bureau to the Nature Conservancy—to put their wish lists to paper within a month. In the years since, more than 1,000 meetings have been held, involving dozens of interested parties, as counties worked on detailed proposals to submit to the offices of Bishop and his congressional colleague, representative Jason Chaffetz, a Republican whose district includes southeastern Utah. Now everyone was waiting to see how the congressmen would merge these often conflicting dreams into one plan without pissing off everybody. Deadlines have come and gone. "It's a total mess," one staffer told me, referring to the difficulty of the process.

Still, it's easy to see why several environmental groups are interested. Over the years Utah's deeply red political culture has often kept the state's enviros on the defensive, simply trying to prevent losses. The Southern Utah Wilderness Alliance, or SUWA, has filed so many lawsuits to stop development on state and federal land that its enemies call it "Sue-Ya." From 1930 to 1980 more than half the lands in Utah that environmentalists would have called wilderness and are managed by the BLM—some 13 million acres—were lost to drilling, roads, and other development. Since 1980, however, greens have held the line. Less than 2 percent of would-be wilderness acres have been lost, thanks to efforts that have included land exchanges, lawsuits against oil and gas drilling, and other tactics, SUWA says.

"Utah has always been hard," says James Morton Turner, author of *The Promise of Wilderness*, a history of modern wilderness politics. "It is the trail that wilderness has never been able to get to the end of. It is steep, straight uphill, hot, and dusty." Utah received no designated wilderness under the 1964 Wilderness Act; even today the state has fewer wilderness acres than Florida. The conflict over land and what to do with it has simmered for decades.

But Bishop's idea strikes a chord with SUWA's 58-year-old executive director, Scott Groene, who sees an opportunity to gain the kinds of permanent protections the group has long sought. A huge chunk of Utah—more than 3 million acres—exists in a limbo state known as wilderness study areas. These places are neither fish nor fowl; Congress has granted WSAs wilderness-like protections while they're studied for permanent status. But some WSAs have been in a holding pattern for upward of 35 years.

Groene believes that haggling to protect lands that already enjoy a degree of protection is "just a waste of our time." Millions of additional acres of unguarded BLM lands are still out there, unprotected even by WSA status, according to environmentalists' surveys. Greens also have their eyes on gaining wilderness protection for some Forest Service lands in eastern Utah. Traditionally the Forest Service's guiding principle has been embodied in the slogan "Land of Many Uses," which means allowing everything from sheep grazing to timber harvests to mountain biking. Wilderness protection, by contrast, is the most stringent status available for federal lands. Motors and mountain bikes are prohibited; resource extraction is allowed only in cases involving preexisting mining claims and leases. Environmentalists have shown a willingness to be flexible, though, by considering less restrictive designations, such as national conservation areas, which block drilling but allow some other uses.

Still, the wish list is long. For example, above Moab environmentalists want wilderness status for the forested La Sal Mountains that serve as the town's watershed, now national forest. To the north they want protection for the Book Cliffs, one of the largest nearly roadless places left in the lower 48. They also want protection for places of staggering beauty and lonesomeness such as Bowknot Bend, where the Green River coils back on itself at the bottom of a canyon of sun-fevered rock. In his most optimistic moments, Groene, a former "hippie lawyer" who now wears dad plaids, allows himself to dream of a day when his group could shut its doors, its mission all but accomplished.

But what would cause the other side to budge?

To look for an answer, I fly into Salt Lake City and drive due east, past the ski resorts of the Wasatch, until the land flattens and bleaches and oil jacks nod their equine heads. This is Uintah County. On the same day Bishop sent his first letter to groups like SUWA, he sent a second, different letter to the elected commissioners of Uintah and every other county in Utah—the level of government that to Bishop best serves the ideal of local control. Bishop acknowledged the long stalemate over their public acreage. Then he told the locals how to break it. Wilderness, he wrote, "can act as currency"—chips that can be cashed in, in exchange for projects like drilling pads, mines, and airports. "The more land we're willing to designate as wilderness, the more we're able to purchase with that currency."

This was new, and it would change the way wilderness skeptics looked at the W-word. The counties of eastern Utah, which had been bloodied and economically frustrated by wilderness stalemates as much as any counties in the state, took notice.

No place illustrates this better than Uintah County. Millions of years ago the Uinta Basin was a vast, shallow inland sea. Today, beneath the ground there's an ocean of hydrocarbons that could make Uintah one of the next global hot spots for energy extraction. The county, which is nearly the size of Connecticut and covers much of the basin, already has more than 7,100 oil and gas wells—and a world-class pollution problem—but that's nothing compared with what boosters hope for. The ground holds huge amounts of additional oil and natural gas, plus oil sands and oil shale. The energy isn't found only in Uintah: in 2012 the most productive oil well in the lower 48 was located outside Moab, and as many as 128 more wells could appear in the area in the next 15 years, the BLM projects.

In Vernal, the Uintah seat, I speak with county commission chairman Mike McKee, an ex-farmer who explains the problem as he sees it. The feds own nearly 60 percent of his county, including many of those potential future oil lands, where would-be drillers are subjected to Washington's bureaucratic red tape, the shifting energy policies of successive presidential administrations, and lawsuits, all of which can make drilling on federal lands more difficult and uncertain than drilling on state lands. If counties like Uintah compromise, officials want dependable assurance that companies really will be able to drill.

McKee stands before a rainbow-colored map in a conference room. It shows the new protections that Uintah County officials think they can stomach and what they want in return. The map is the outcome of dozens of meetings involving everyone from snowmobile groups to the Wilderness Society, McKee explains, and similar maps have been cobbled together all over eastern Utah. Up in one corner there's a new wilderness area in the Uinta Mountains. Another is marked in the south, for Desolation Canyon. ("Their ultimate prize," McKee says of environmentalists. "A spectacular area.") Negotiations are ongoing about the possibility of turning Dinosaur National Monument into a national park. And a blob of gray covers roughly half the map—a proposed new "energy zone," where current or future oil or gas drilling or tar-sands operations could occur.

It might be hard to find enough common ground to facilitate any huge deals simply by swapping lands. But Utah has something else to grease the bargaining wheels. During the statehood processes that occurred when the West was settled, the federal government gave new states land, arranged in a checkerboard pattern, most of it to be logged, mined, or grazed, with the proceeds largely helping to fund state schools. In Utah today these trust lands make up 6.5 percent of the entire state—millions of acres. But since they are still arranged in checkerboard patterns, many sit inside federal lands that are now wilderness study areas; they're like holes in a doughnut. Critics, including Bishop, say this effectively makes the land undevelopable, leading to underfunded schools. (The real cause of that problem, many believe, has been Utah's tax-averse citizenry.) However, these state lands can be legally mined and drilled, and the federal government has to allow access to them, even through potential wilderness. That's why Groene calls them "a ticking bomb."

Many on both sides would like to see them on the trading block. Environmentalists want the development threats removed. Counties and state legislators want the parcels traded out for other lands that are more accessible. Both sides have something to gain.

During the time I spend researching the grand bargain in Utah, many people tell me that what is unfolding might represent the future of land conservation in America. "This could be a model for other states to resolve these issues in a constructive way," says Mike Matz, who directs wilderness conservation projects for the Pew Charitable Trusts, one of the nation's largest conservation groups. Utah's collaboration has spurred counties in Wyoming to try to settle the future of hundreds of thousands of WSA acres there. Whether this trend should be applauded is hotly debated, however.

Land conservation in the U.S. is harder today than ever. The easiest places to protect—the high country of peaks, pines, and pikas—have already been taken care of. What generally remains are lower-elevation lands that have more claims on them, whether from off-road drivers, mountain bikers, or energy companies, according to Martin Nie, director of the Bolle Center for People and Forests at the University of Montana. Add a highly polarized Congress that won't pass anything even remotely controversial and you've got a formula for inaction. "The market is dictating that tactics need to change," says Alan Rowsome, the Wilderness Society's senior government-relations director for wildlands designation.

Enter collaborative deal-making. Collaborations have their roots in the watershed groups that formed in the late 1990s, during the Clinton administration. Collaboration often comes down to stakeholders—ranchers, loggers, ATV groups, environmentalists—sitting in a room and negotiating about lines on a map. The goal is to strike a balance between the needs of nature and the ever-greater demands of people. Over the past 15 years some of the nation's largest and most respected environmental groups— including Pew, the Wilderness Society, and the Sierra Club—have played active roles in collaborations around the West. At its best, collaboration has helped break logjams in stubborn public-lands conflicts and has created a new way forward to protect significant chunks of the map.

But some environmental watchdogs, wilderness specialists, and academics worry that the approach is also setting dangerous precedents. In their pursuit of land preservation and wilderness, critics charge, environmental groups frequently horse-trade inappropriately with the public's lands—shutting out dissent, undercutting their conservation mission, and even eroding bedrock environmental laws.

Many say the new collaborative trend began in a remote place in eastern Oregon called Steens Mountain. In 2000 environmentalists and ranchers brokered a complex agreement that involved new wilderness and land exchanges and also created a novel management area on public land that would be overseen in part by the ranchers themselves. Some environmentalists praised the compromises, but to Janine Blaeloch, founder and director of the watchdog Western Lands Project, the deal smelled funny. Blaeloch was particularly concerned about how conservationists gained protections around Steens only if they "paid" for them with outright losses of other public lands or by agreeing to reduced protections elsewhere. Blaeloch dubbed it "quid pro quo wilderness."

The trend soon gained popularity in Nevada, where four-fifths of the state is managed by the federal government. Having had success with a deal in Clark County, which surrounds fast-growing Las Vegas, Democratic senator Harry Reid tackled a more collaborative effort in a large county to the northeast, Lincoln. At first glance the 2004 Lincoln County Conservation, Recreation, and Development Act seemed like a success: it created 14 wilderness areas covering nearly 770,000 acres. The bill had the support of the Wilderness Society and the Nevada Wilderness Coalition, the latter supported financially by a Pew-backed group called the Wilderness Support Center.

But a closer look revealed a disturbing deal. Nowhere did the Lincoln County bill break more with the past than by going against a long-standing government policy of trying to keep federal lands in federal hands. Instead, according to historian James Morton Turner, it required the feds to sell more than 103,000 acres of public land at auction.

Twenty years ago, when environmentalists haggled over protecting lands, they sometimes compromised in a way that temporarily returned hoped-for wilderness acres to the kitty while they continued fighting to protect them at some future date. In Nevada the environmental groups hadn't simply delayed protection for public lands; they allowed them to be sold off entirely.

Inspired by Lincoln County, other collaborative efforts followed, including a controversial 2008 deal across the state line in southwest Utah's Washington County and another in the dry Owyhee country of southwest Idaho in 2009. And why not? "Collaboration" sounds great. It suggests consensus and compromise—the idea that everyone will be heard and their ideas made part of the finished product. But as George Nickas, executive director of Wilderness Watch, has said, compromise sometimes means "three wolves and a sheep talking about what's for dinner."

In short, whether collaboration is a good thing or not depends a lot on where you stand—and what you stand to gain. A 2013 study found that the groups most likely to collaborate are large, professional environmental organizations that often represent diverse agendas. According to Caitlin Burke, a forestry expert in North Carolina who has studied collaborations, if such trends continue, "we will see a marginalization of smaller, ideologically pure environmental groups [whose] values will not be included in decision-making because they are unable or unwilling to collaborate."

Idaho's Owyhee Initiative, for instance, which steers issues like ranching in southwest Idaho, has banned from participation groups that have successfully litigated to reduce grazing in areas of the Owyhee. The initiative, however, preserves a role for more mainstream groups like the Nature Conservancy and the Wilderness Society.

Despite appearances, collaborations are undemocratic, argue critics like Gary Macfarlane of Friends of the Clearwater, an environmental group in northern Idaho. The public already has a process for how changes can be made to our public lands, Macfarlane says: the 1969 National Environmental Policy Act. Macfarlane describes it as "a law that tells federal agencies to look before you leap" and says you have to allow all interested parties to participate. The act also mandates that the best available science be considered. Collaborations don't have to do that, says Randi Spivak, director of the public-lands program for the Center for Biological Diversity.

Then there are the concerns about wilderness. Designation of new wilderness areas has often been a centerpiece of collaborations over the past 15 years. But in order to push wilderness through, the big environmental groups have been willing to make sometimes disturbing compromises, critics say—even to the Wilderness Act itself.

Compromise has long been a central part of wilderness politics, of course. The 1964 Wilderness Act took eight years and 65 bills to become law, and the final act grandfathered in some grazing and mining. But the old compromises were largely about boundaries —what's in and what's out. The new deals embrace a more insidious type of compromise, not just about where wilderness will be but also about how it will be managed.

"Our fear is that some conservation groups look at the 1964 act as the place to begin a new round of compromises," says Martin Nie. That shift, he adds, "could threaten the integrity of the system."

In collaborative efforts, large conservation groups that badly want to protect wilderness must deal with groups that sometimes loathe the idea, so conservationists increasingly feel pressure to make wilderness more palatable to opponents—and that means watering it down, says critic Chris Barns, a longtime wilderness expert who recently retired from the BLM.

The number of special provisions—exceptions added to a wilderness bill, almost always leading to more human impact—has increased in the past several years, according to a 2010 study in the *International Journal of Wilderness*. The Lincoln County deal was saddled with a raft of such provisions. The Owyhee deal, given a thumbs-up by such groups as Pew and the Wilderness Society, lets ranchers corral cattle using motorized vehicles, which is supposed to be forbidden in wilderness. The result of such compromises, Barns and others say, is areas known as WINOs—"wilderness in name only."

Another problem with these exceptions is that they become boilerplate for future bills, Barns says. A provision that first appeared in 1980 has since turned up in more than two dozen wilderness laws. Such changes might seem small, says Barns, but they erode, bit by bit, America's last wild places.

The large environmental groups involved in these deals say such criticisms are off-base. "For us, what it comes down to is, as the nation grows, there are less and less wild places left in America," says Alan Rowsome of the Wilderness Society. "And we want to protect them as soon as possible, because if you wait 10 years or 15 years or 20 years, that place may not be protectable. It may not be as wild a place, or as open, or as important to protect."

As for the criticisms about horse trading, Pew's Mike Matz says, "I would submit that these deals have always been complex. At the end of the day, we are entirely comfortable with the deals we have struck."

But Athan Manuel, lead lobbyist for the Sierra Club on federal-lands issues, acknowledges some of the criticisms. "We don't have to get some of these bad deals just because we think the atmosphere is going to be worse next year," Manuel says. "I think we sometimes suffer a crisis of confidence in the environmental community."

Which brings us back to Utah and the grand bargain. With Pew and the Wilderness Society heavily involved throughout much of the process, critics were watching and worrying. Though Bishop hadn't yet unveiled his final proposal, some didn't like what they were hearing—that green groups were willing to swap out Utah's school trust lands for parcels located elsewhere, so that wildlands could be consolidated and the state could more easily drill for oil and gas. "It's wilderness for warming," Randi Spivak says. "We should be keeping fossil fuels in the ground."

"It's painful," SUWA's Groene acknowledges, "because everything you give to Utah you have to assume will be sacrificed." But the kinds of ethical dilemmas that collaborations have posed elsewhere? He's still waiting on them. "The whole goal was to be put in that terrible place where we had to decide" those kind of things, he says. But they aren't even there yet.

When I asked people for a place that exemplified the challenges and promise of the grand bargain, many pointed to San Juan County, which anchors Utah's bottom right corner. Though larger than New Jersey, it has only 15,000 residents, many of them conservative sons of Mormon pioneers.

The current county commission chairman there is a man named Phil Lyman, who for the past three years has spearheaded the county's public-lands advisory council, a citizens' group that is charged with helping officials craft a proposal to give to Bishop. In the midst of that work, Lyman was convicted of leading an illegal ATV excursion into Recapture Canyon, an area that had been partly closed to protect ancient Native American ruins.

Lyman lives in Blanding—at 3,600 people, the county's largest town—which was founded by his great-grandfather in the 1890s. It has a bowling alley, a bank, and a few wide streets that seem designed to take tourists elsewhere. On Main Street old ranchers limp in and out of the San Juan Pharmacy. Down the block I find the door to Lyman's accounting office open. A sign inside invites visitors to take a complimentary copy of the Book of Mormon.

To environmentalists, Lyman, 51, symbolizes all that is backward in Utah. To others he's a hero for resisting government overreach. The day I meet him, he's still waiting to find out if he'll go to prison over his conviction for conspiracy and trespassing. (He was already fined and in mid-December would be sentenced to 10 days in jail.) He scratches his head at the furor he has caused. It was all a misunderstanding, he claims.

"I'm the poster child for gun totin', beer drinkin', ATV drivin'," he says. "I'm an accountant!" He doesn't even own an ATV. "But I do have a fire in the belly about hurting innocent people," he says.

We drive into the sagebrush desert in Lyman's pickup, and he tells the same story I've heard elsewhere about how reluctant he was at first for his county to join the grand bargain. All it would do is stir up trouble for years, with no resolution—and the county was already trying to start talks with a Navajo group. But once he decided to get on board, swayed by his elected duty to give the approach a try, Lyman threw himself into it.

To everyone's surprise, people with different viewpoints actually started talking to each other, Lyman says. Around eastern Utah, cooperation was in the air. The Wilderness Society joined the grand bargain early on. So did Pew, though it generally pursued its own track. In the spirit of mediation, greens told Bishop that in some places they would be willing to accept less proscriptive national conservation areas instead of a stricter wilderness designation, which many counties feared.

Working county by county, the collaborators nailed down the first tentative deal—a single puzzle piece in the larger grand bargain—in the fall of 2014 in Daggett County, a smaller county in northeastern Utah. (Individual deals like this, the hope went, would eventually be submitted to Congress as one overarching Utah lands bill.) Among other things, the agreement designated over 100,000 acres of wilderness and conservation areas in exchange for the acquisition of a large natural-gas facility and future development of a ski area.

Momentum was building. Summit County, home to Park City, asked to contribute to the Public Lands Initiative. An election in Grand County, home to Moab, saw ultraconservative commissioners voted out and a more receptive slate voted in. At various times as many as eight counties across eastern Utah were in play.

But just as the grand bargain looked most promising, the wheels started to come off. One month after Daggett County inked its deal, elections swept very conservative candidates back into power there. The new county commission reneged on the agreement that had just been celebrated. As Groene sees it, Bishop had made a crucial error by telling county officials that they had the power to stop any negotiation if they didn't like how it was going. If you're a small-town politician in rural Utah and you strike a deal that makes a lot of concessions to environmentalists, he says, "then all the people you know are going to yell at you in the grocery store the rest of your life. Politically they can't play that role in these small towns."

When Daggett County reneged, Bishop didn't seem to have a Plan B to keep momentum going elsewhere, Groene says. Bishop calls this criticism "terribly unfair." After all, he says, "anyone can walk away" from this process—even the environmental groups.

Last summer Lyman's San Juan County was the last of seven counties to hand in its blueprint to Bishop. It called for 945,000 acres of protection, a mix of wilderness and national conservation areas. Parts of famous geologic landmarks such as Cedar Mesa, Comb Ridge, and Indian Creek, with its world-class rock climbing, would get bolstered protection. It was far more than many people expected—or than many others could tolerate.

"No one would have predicted that San Juan County would've come up with a proposal that included a million acres of land protection, and part of that protection uses the W-word," says Josh Ewing, director of the group Friends of Cedar Mesa.

Yet the proposal still needs a lot of work, conservationists say. For one thing, the county's plan (along with several others') contains language inoculating it from the Antiquities Act. This law gives the president the power to preserve landscapes by creating national monuments on federal land without requesting local input or consulting Congress. Because San Juan County and other eastern Utah counties are more than 60 percent owned by the feds, many state politicians hate the act, calling it the epitome of government overreach. While the county sees the exemption as insurance that its hard work won't result in a monument, conservationists consider an exemption a deal-breaker.

Then there's the Native American issue. About half of San Juan County's residents are Native Americans, and about one-quarter of the county is reservation land. Because of a general lack of outreach, but also by their own choosing, only a few Native Americans participated in the county's grand-bargain planning. "It's a trust thing," Lyman acknowledges. "I don't blame any Navajo personally who doesn't trust the white community, the federal community, the county."

Lyman drives up Highway 95, the road where Seldom Seen Smith yanks survey stakes in Edward Abbey's 1975 novel *The Monkey Wrench Gang*, and turns onto a dirt road. He stops the truck among

Mormon tea and sagebrush, where the earth is slowly reclaiming the ruin of an old uranium mill. Bones of rebar poke through concrete. Lyman tells me that his grandfather once owned the nearby uranium mine, one of the richest around.

Conservationists want protection here. But under the county's proposal, this would be an energy zone. Right now, Lyman says, there isn't a mine operating anywhere from Durango to Hanksville. "Let's really draw the line," he says. "West of here would be the wilderness they want, and east of here would be the really productive public lands, managed intelligently. Real people, good jobs, putting food on the table."

He toes a loose bit of concrete and squints into the hammered brightness of a desert noon. "When you're from here and your ancestors were part of this," he says, grabbing his chest, "you ask yourself, What am I doing? What's our legacy? Do we have anything to leave and show we were effective?"

Is this place God's cathedral or man's quarry? I'm reminded that how you answer depends on where you stand and where your grandfather stood before you.

On my other morning in San Juan County, I head out with Josh Ewing to see an alternate future for this place. After a 90-minute bounce over rough roads, Ewing stops his truck and disappears among old-growth juniper. I jog after him. We're atop Cedar Mesa, a massive high plateau whose edges are fissured with canyons and that once was the northwestern frontier of a prehistoric civilization, home to the ancestors of today's Hopi, Zuni, and others. Ewing—bespectacled, fast-talking, a climber who fell hard for the red-rock country and is now director of Friends of Cedar Mesa—hustles through the sage and sand. "It gets in your bones," he says as we walk past yucca and prickly pear, ripe with purple fruit. "It's got that beauty combined with the archaeology."

The junipers part, and the earth yawns open in a canyon of tall sandstone walls striped like candy. High on one wall perch several ancient cliff dwellings. We follow steps hacked into the stone that lead to ledges—the ground falling away 500 feet—until we reach walls, stone rooms, a granary. In the dimness of a kiva, or ceremonial room, a half-dozen gnawed corncobs stand upright as if in an artist's still life. Ewing guesses the place is 800 years old.

Ewing shakes his head: this side canyon and its relics aren't protected in the plan San Juan County submitted to Bishop, he says, and they won't survive another millennium without help. "Things don't stay the way they are by not doing anything," he says. Here was that difference again: while Lyman wanted to leave a legacy by developing the land, Ewing and other conservationists wanted to do so by protecting it. His group estimates that there are more than 100,000 archaeological sites in the greater Cedar Mesa area — cliff dwellings, burial sites, kivas, petroglyphs, and pictographs. "There are probably closer to a quarter million."

I tell Ewing how all that week in places like Moab environmentalists said that the grand bargain was already on life support. The counties' proposals were totally unrealistic, they said. Even Pew recently stepped back from talks, saying it couldn't abide an Antiquities Act exemption. Ewing is more optimistic.

"I don't think a legitimate negotiation has started yet," he says. Only after Bishop unveils his draft proposal, he predicts, will the real, painful tradeoffs happen. Why would he force his conservative constituents off their positions any sooner?

And if that movement doesn't happen? I ask Ewing.

Then this play is over, he replies, at least as far as the legislative effort is concerned.

At the cliff's edge we talk about the possibility of a Plan B. For years conservationists have been working two tracks, also prodding President Obama to declare a massive national monument in eastern Utah before he leaves office. The prospect infuriates many rural Utahans, who still feel stung by Bill Clinton's creation of Grand Staircase–Escalante National Monument in 1996. The pressure has given an urgency to the process.

Ewing sweeps an arm before us to introduce the leading monument candidate: the Bears Ears, a 1.9-million-acre triangle of land that would encompass all of Cedar Mesa, including more than 40 mini–Grand Canyon systems that few people ever visit. (The name comes from the twin rock towers that rise over the landscape.) Twenty-five Native American tribes, some of them historic enemies, have united to urge Obama to create the monument.

As SUWA's Groene sees it, the old battle lines are being drawn again, pitting cowboys against Native Americans and environmental groups. On one side are the counties, supporting a flawed grand bargain; on the other, tribes and greens pushing a monument that county officials hate. This time, however, the Great White Father in Washington, D.C., is black. He has declared or expanded more national monuments than any previous president and, sources tell me, is ready to give an underserved people what they want. (The White House did not respond to requests for comment.)

The night before I leave San Juan County, I have dinner in the village of Bluff with Mark Maryboy, the first Navajo to serve as a county commissioner there. Maryboy is as tall and lean as a piece of jerky, and he wears the pointed boots of a former rodeo bronc rider. I ask him why the Bears Ears is so important to Native Americans.

Maryboy doesn't answer directly. Instead he tells stories. He says his grandparents used to live on Cedar Mesa. He says the Navajo have stories about everything—about the horned toads on the mesa, about how the sun is called Grandpa, about the male and female winds that live just north of the Bears Ears and how when a person has a breathing problem he will seek out the right one to heal it. There are hundreds of stories, he says.

Only after I leave him do I understand what he meant. He wasn't talking about acres. He was talking about a way to live.

On January 20, Bishop and Chaffetz unveiled their draft proposal. It would give added protections to about 4.3 million acres of Utah, roughly split among 41 new wilderness areas and 14 national conservation areas, as well as seven "special management areas." To try to address Native Americans' concerns, the proposal would create a nearly 1.2-million-acre conservation area for the Bears Ears, to be managed by federal agencies and advised by a commission containing some native people. It would expand Arches National Park and add more than 300 miles of wild and scenic river protection.

On the other side of the ledger, the proposal would release nearly 81,000 acres of WSAs, meaning that they could no longer be considered for wilderness status. It would create large zones for energy development. It would hand over some 40,000 acres of federal land to state and local governments. And it would give Utah right of way to a large majority of those contested 36,000 miles of dirt tracks that spiderweb eastern Utah's canyon country, often on federal land—clearing the way for the state to open them to off-road vehicles, or improve them, or even pave them, despite the wishes of Uncle Sam. As for the Antiquities Act—that part, curiously, was not addressed, so the counties will be able to suggest their own wording. The final bill "must include language that guarantees long-term land use certainty," Bishop's office wrote in a summary of the proposal.

Everyone was destined to be unhappy with aspects of the draft, Lynn Jackson, a Grand County commissioner, tells me, adding, "This is what compromise looks like."

It may also be what failure to achieve compromise looks like, because reviews by environmentalists were rapid and negative.

To the eyes of SUWA's Groene, the deal would give the farm to counties, off-road users, energy companies, and the state of Utah in exchange for little sacrifice on their part. "Those are swell-sounding numbers," he says of the wilderness acreage, "but this whole thing falls like a house of cards when you look at what those numbers mean on the ground." Several of the new wilderness areas, for instance, would lie within existing national parks, where the lands already enjoy significant protections. Taken together, Groene says, the proposal "actually means less protection than currently exists, advancing the state of Utah's quest to seize our public lands and igniting a carbon bomb."

So what now? Everyone will demand changes. The congressmen will take those demands and reemerge with a final proposal. "My name is on it," Bishop told me last fall. "I'm not playing games with it. I'm not putting in stuff that I'd be willing to barter later." The timeline for all this? Bishop seemed eager to keep things moving. He was clearly exasperated with the grinding process.

But environmentalists are no longer in the mood to play ball. "The bill is unacceptable and unsalvageable," Groene says. "If they'd be willing to have a do-over, we would be willing to have discussions." In San Juan County, Groene's group won't negotiate at all—they want the Bears Ears National Monument.

Another significant defection happened three weeks before Bishop's draft appeared, when the Bears Ears Inter-Tribal Coalition announced that it was formally withdrawing from the Public Lands Initiative and throwing all its efforts into lobbying Obama for a national monument, citing months of missed deadlines, delays, and "no substantive engagement" with its concerns. "We don't feel we can wait any longer," the group said.

Weeks earlier, before all this, I asked Bishop if he was still optimistic about his grand bargain. "If it's an even-numbered day, I feel positive," he quipped. The Obama administration had been encouraging, he said. "They have never given me a deadline for anything per se."

It was hopeful for Bishop to think that after so much work, the proverbial win-win was still within reach. Now that seemed more unlikely than ever. The Native Americans had been alienated. And despite Bishop's assurances, time was running short to push anything through Congress during an election year, even with the support of powerful Utah senator Orrin Hatch.

The only certainty was that even the players on this stage didn't know what was coming next—whether it was the end of Utah's wilderness wars or just the close of yet another dyspeptic chapter.

"Make sure you've got popcorn," said Casey Snider, Bishop's legislative director. "It's going to be quite a show."

PART III

The "Real Life" of Scientists

SALLY DAVIES

The Physics Pioneer Who Walked Away from It All

FROM *Nautilus*

INSIDE THE SOUTH London offices of Doppel, a wearable technology startup, sandwiched into a single room on a floor between a Swedish coffee shop and a wig-making studio, CEO and quantum physicist Fotini Markopoulou is debating the best way to describe an off-switch.

Markopoulou and her three cofounders have gathered in convivial discomfort around a cluttered Formica table and lean-to blackboard. They're redesigning the features of their eponymous first device, which is due to be released in October. It's a kind of elegant watch that sits on the inside of your wrist and delivers a regular, vibrating pulse. By mimicking a heartbeat, the Doppel helps regulate a person's emotions and mental focus.

Swiveling in a chair, Markopoulou says she likes a "smothering" gesture — placing a palm over the face of the Doppel to turn it off — because it is intuitive and simple, and the term suggests the device is "alive." "You could always murder it," deadpans commercial director Jack Hooper. Head of technology Andreas Bilicki chimes in: "Why not *choke* or *asphyxiate*?" The team throws around alternatives: "throttle"; "go to sleep, to sleep"; "turn your Doppel off, just like putting a blanket over a parrot's cage."

Markopoulou, 45, observes the banter with a half-smile. She is fine-featured and striking. Her heavy-lidded eyes anchor a gaze that seems wary of its own powers, as if her promiscuous intelligence must hold itself back from latching on to your every word.

She wears her hair in a tousled pixie cut and, on this spring day, a green knit sweater and blue scarf with a pattern of fishlike scales. There are no airs about her, nor any indication that she's 20 years older than the rest of the team. Markopoulou lives in Oxford but sleeps on design director Nell Bennett's couch whenever she comes down to London.

After the meeting Markopoulou and I walk downstairs to get a coffee. With the zeal of the reborn, she tells me how much she relishes the pleasures of making a product that people will use and pay for. "There is a very practical satisfaction to getting stuff done, whether it's making something or selling something," she says. "I do enjoy solving practical problems, like how to convince people Doppel's a good idea, or how to get the right deal from an accountant."

It's hard to see how these tasks could fully absorb Markopoulou. She is one of the most radical and fiercely creative theoretical physicists alive today, and a founding faculty member of the Perimeter Institute for Theoretical Physics in Waterloo, Canada, where she was at the vanguard of quantum gravity. This is the branch of physics striving to unify the two most fundamental theories of the universe: general relativity, proposed by Einstein, and quantum mechanics.

Quantum theory describes the rowdy interactions of fundamental particles that govern many of the forces in the known universe —except gravity. Gravity is rendered beautifully predictable by general relativity, which envisions it as an effect of how the four dimensions of space and time curve in response to matter, like a piece of tarpaulin bending under a bowling ball. Quantum theory's ability to predict the behavior of an electron in a magnetic field has been described as the most precisely tested phenomenon in the history of science. But putting it together with gravity has so far produced absurd mathematical results. It's as if a soccer player and a tennis player were managing to carry on a game despite being ignorant of the opponent's rules.

After years of single-minded study, Markopoulou cocreated a novel potential solution known as "quantum graphity." This model of the universe operates at a scale that is tiny even by subatomic standards—as tiny in relation to a speck of dust as a speck of dust is to the entire universe. It suggests that space itself and its attendant laws and features could evolve out of interconnected dots to create the dimensions we experience as space, like a soufflé rising from a pan.

"Fotini is extremely original, original to a fault," says Lee Smolin, a fellow founder of Perimeter who used to be married to Markopoulou. "Most scientists pick up on ideas which are dominant, which come from living figures, and develop them incrementally. She doesn't do that—she works solely on her own ideas."

Between sips of a latte, Markopoulou describes how theoretical physics consumed her. "It's a lot like being in a monastery, like no normal human needs should make you waver from the cause of understanding where the universe came from," she says. "In my previous eyes, just leaving is a moral failure, more than anything else. It's a devotion thing—your devotion has just gone." She pauses to shape her next thought. "It's also not really a loss of faith; I changed."

Five years after walking away from physics, Markopoulou is still trying to explain that change to herself. She was forced to reexamine her position when Perimeter's new director, Neil Turok, who joined in 2008, deemed her work too speculative and squeezed her out of the institute. But her unease had deeper roots.

Working in a field where the air of reality was so thin, Markopoulou started to lose touch with her own life. "I have so many friends in their late 40s, and they still don't have an actual home or a family or anything. As long as they have a place where they can go and think, they're happy." She casts a wry smile. "I failed that test, obviously. For a lot of people that makes sense, and even for me that makes sense 80 percent. It's that other 20 percent that causes problems."

Doppel embodies many of the qualities that Markopoulou came to miss in her work as a physicist. The company draws on the science of psychophysiology, a field which considers the mind to be deeply rooted in the body and its environment. But embracing the fact that the self is interwoven with the world, and at its mercy, is a frightening thought, Markopoulou says. Escaping that fear, and trying to pin down the interconnection between humans and the natural systems that make us what we are, is what drew her to physics in the first place.

"I did appreciate, for a long time, the way science detaches you from that scariness, because you ignore it," Markopoulou says. "Between the truth of the physical world and a physics theory, there's humans. Of course, nothing happens there, because removing the person is the whole point of training as a scientist." A pause. "But this may or may not be possible."

As a teenager growing up in Athens, Greece, Markopoulou looked like an ordinary kid: permed hair, heavy ribbed sweaters, a penchant for Clint Eastwood Westerns. But she was already attracted to the study of transcendent truths. On her way home from school, she would sometimes drop in at Greek Orthodox churches to lie on her back in the pews and contemplate the elaborate scenes of stars and angels painted into the interior domes. One summer, when she was 15, she happened across a book in the library of the British Council with the title *Starseekers,* a quasi-mystical account of the history of cosmology by English writer Colin Wilson. "I got totally obsessed with that book," Markopoulou says. She convinced her mother, Maria, to buy her an Atari computer, and spent hours trying to translate *Starseekers* into Greek on a word processor.

Markopoulou lived with her mother in a cramped two-story studio in Athens, where Maria worked as a figurative sculptor. She was a magnetic, troubled figure, unafraid to set her own moral compass but riven with internal conflicts. She'd fallen pregnant one summer to a Greek sculptor she had known in Florence, where she trained as an artist against the wishes of her parents, and decided to raise the baby as a single mother in Athens. She was 33. "Her lovely way of putting this was, 'Jesus was also 33 when he was put on the cross,'" Markopoulou says. "But it was also very clear that I was the best thing that had ever happened to her." (Markopoulou has never met and knows little about her father, who died in 1997.)

Markopoulou loved accompanying Maria to exhibitions and openings but struggled to disentangle her sense of self from her mother's strong and particular judgments. "My mother's relation with reality, it would be wrong to say that it wasn't solid, but it was just different," Markopoulou says. Maria hated to sleep and refused to have a bed: "My mother clearly thought that sleeping was like dying, and that she might not wake up if she did, and something like a bed might as well have been a tombstone. I did realize as I was growing up that you couldn't rely on her description of something."

The subjectivity of aesthetic merit troubled Markopoulou. "One of the things I hated about the art world is that decision-making is quite arbitrary," she says. "People could say Picasso is shit just because they felt like saying it. I found that very frustrating, and very political; they're gatekeepers, and then your life and self-perception is a function of those gatekeepers."

Markopoulou's education in Greece was "a complete disaster," she says, with teachers whose instruction consisted of reading the newspaper at the front of the classroom. In her final year of high school, Markopoulou went in search of private evening classes; by mistake she walked through the door of an institution that offered A levels, the exams for students entering the British university system. She hadn't considered studying in the United Kingdom, but ended up enrolling. "The usual story about people in quantum gravity is, 'I read about Einstein when I was eight,'" she says. That was not her. The pendulum for her imaginary career had swung between being an astronaut and an archaeologist. She only selected theoretical physics under the pressure of her university application, and chose the course on the casual advice of a tutor at the school, a former NASA scientist, who said it would be a good balance for her aptitude in physics and mathematics.

Markopoulou says she failed her A levels—"the first time I walked into the lab was for the exam, and half the questions I answered in Greek"—but, as part of the clearing process between teachers and universities, her tutor secured her a place at Queen Mary's University in London, her first choice. The department had several excellent particle physicists investigating the top quark, but the place retained the welcoming atmosphere of institutions unburdened by hallowed reputations.

Money was tight, so Markopoulou didn't have much of a social life. She planned birthday parties at McDonald's for a bit of extra cash, while her mother, who was living with her in London to get the rent from her studio in Athens, repaired antiques. (They would continue to live together until the last year of Markopoulou's PhD.) But Markopoulou loved it all the same. She and a clutch of the other undergraduates would relax in the chapel café between lectures and occasionally head out in the evenings to hear one of their professors play amateur hard rock. At the same time, in her classes she got wind of the fact that "there was some forbidden place"—that when it came to certain subjects, such as why time moves in one direction, it was better not to ask. She was not

content with what the rules were; she wanted to know how they
came to be.

Toward the end of her undergraduate degree, a friend suggested
Markopoulou attend a lecture on quantum gravity by Chris Isham,
a rigorously mathematical physicist at Imperial College. He was
also a Jungian analyst and devout Christian, with the air of a mystic
and a fondness for peppering his lectures with passages from T. S.
Eliot and Heidegger. "You can't take out of the world the fact we
see it," Isham tells me. "What is the reality we hang on to? Well, it's
us, but who are we that sit inside this space which is relative to us?"

Isham was the first person Markopoulou encountered who
could relate the technical dimensions of science to humans' wider
search for meaning. "Sometimes doing physics can be a bit like
doing plumbing—you have your equations and tools and you go
around and fix stuff, and if you do it in a smart way, people respect
you," she says. "Because you are a professional physicist, you get
used to the idea that there are difficult questions that you do not
do for a living. But these are what drove most of us to join the
ranks."

Markopoulou was developing her own clear vision of what she
wanted to achieve as a physicist. "I am not going to devote my life
to something because it's beautiful—it's this quest for the truth,"
she says. "Science is not philosophy—there is not a lot of value in
thinking about questions if you cannot come up with answers. But
I've always been attracted to what is the furthest away you can get
such that you can still come back with an answer. You're trying to
find the end of the coil to unfold it."

Under Isham's influence, Markopoulou started to grapple with
quantum gravity. Her assigned PhD project was based on a previous paper that examined the movement of dust particles to develop a new approach to splitting time away from the three dimensions of space. This sounds like a solution in search of a problem
—surely time is a different thing from space?—until you remember Einstein's counterintuitive insight that time is intimately interwoven with the fabric of space and can be similarly twisted and
bent by matter and movement. Time is dynamic, and defined by its
relationship to what's happening around it. It follows that there is
no absolute time that the whole universe obeys—and, more trou-
blingly, when you push the equations far enough, time has a ten-

dency to disappear entirely. "The relativity view of the world is that space and time is out there and it's more or less a static thing — time is just another dimension," the distinguished physicist Roger Penrose explains to me.

However, Einstein's account of time doesn't make sense in quantum theory. The quantum realm is host to all sorts of phenomena — particles existing in two places at once, or becoming entangled, as if they're able to communicate their properties instantly and seemingly telepathically, whether separated by a lab bench or a light-year. It adopts a version of time that's far more conventional, like a metronome ticking away in the background, distinct from the bizarre behavior of quantum theory's zoo of quarks, bosons, and fermions.

It began to dawn on Markopoulou that you might be able to reconcile these two accounts of time by looking more closely at how they viewed space. After her first paper on dust modeling, she turned to spin networks. These are geometric models which help physicists describe quantum interactions in space and fit more readily with the mathematics of general relativity. Markopoulou had the idea of combining spin networks with a "causal set," which allows time to be captured as a history of discrete events rather than a continuous flow. Showing how histories could be represented spatially let her bring a more substantive version of time into general relativity — one that wasn't rigid (as in some accounts of quantum theory) nor completely flexible (as in general relativistic spacetime).

Her work caught the eye of Smolin, an American theoretical physicist who at the time was visiting Imperial from Penn State University. He'd made a name for himself as a joint inventor of loop quantum gravity theory — a competitor to string theory in the quantum gravity sweepstakes — which was building on spin networks to develop a more sophisticated picture of quantum spacetime. Smolin worked with Markopoulou on a paper on causal sets, and invited her back to Penn State for three months while she was finishing her dissertation. They went on to marry in 1999.

At the time Penn State was a premier institution for nonstring quantum gravity, and Markopoulou was surrounded by a number of other brilliant young scientists. "A bunch of different ideas were coming together; there was this sense that you might actually do something faster than the person in the next room, which is very unusual in quantum gravity," she says. String theory had never appealed to Markopoulou, who saw it as cutthroat and conformist. "String theory has a very strong pecking order," she says. "It comes with a strong machismo: What complicated stuff can you do? They're very good at maintaining that."

Some of Markopoulou's contemporaries saw the equations pointing to the conclusion that time is an illusion at the fundamental level, and that what we experience as the progression of events emerges as a byproduct of fluctuations in space. But Markopoulou tended to attack the problem from the other direction —looking at time as the most important thing and space as something that grows out of it, or is left as a trace, like a logbook of what has taken place in time. "I'm a bit extreme in that I would actually like to keep a fairly old-style time," Markopoulou says. "I'm not wrong in my views. They come with challenges, but they also come with opportunities."

In early 2000 whispers went around the theoretical physics community that somebody wanted to donate $100 million for an institute dedicated to foundational physics. Markopoulou and Smolin were approached by Howard Burton, a Canadian with a PhD in theoretical physics from the University of Waterloo, who was the emissary for this Gatsby-like figure. "I genuinely thought the guy was a sociologist studying the reactions of physicists to that statement—the amount of money is crazy for foundational physics," Markopoulou says. By now she was doing a postdoctoral fellowship at the Max Planck Institute for Gravitational Physics in Berlin. She and Smolin were flown in secrecy to Canada and only informed that the donor was Mike Lazaridis, the founder of BlackBerry, on the drive from the airport: "We spent the night at Mike's house, where he made us French toast and talked us into coming to Waterloo."

By the time Perimeter was set up, she and Smolin had separated, but remained friends—which was just as well. To lay the scientific groundwork for the institute, the three founding faculty huddled together in a former restaurant, along with several postdoctoral students and Burton, whom Lazaridis had appointed as the director. They inherited the coffee machine and learned to make top-notch barista cappuccinos. The institute aspired to a flat management structure hospitable to freethinkers, without tenured jobs or the ordinary hierarchies of a university physics department, in the hope that this would foster more interesting research.

While Markopoulou was not a forceful leader, she was a persuasive one, says Seth Lloyd, a physicist and professor at the Massachusetts Institute of Technology and a longtime collaborator of Markopoulou's. He recalls trekking with her and some postdocs in the Sangre de Cristo Mountains when she was on a fellowship at the Santa Fe Institute in New Mexico. "At each stage of the hike there were different suggestions about where to go, and we always ended up doing what Fotini thought was a good thing," he says. "We had a great time, none of us ever thought Fotini was imposing her will—just that what Fotini seemed to want to do was the right thing to do."

At Perimeter, Markopoulou was at her best when the learning and experimentation were the quickest. Invariably her work became playful and synthetic. "At some point I thought we should just reduce the whole thing to the basic property of space, which is here and there," she says. Physicists were willing to toy with the nature of time and "hack" general relativity to create a quantum gravity theory, she says. But they seldom played with the nature of space or interfered with quantum theory. With Simone Severini, an Italian computer scientist, and graduate student Tomasz Konopka, Markopoulou drew on quantum information theory to develop the notion of quantum graphity. "Fotini thought it was fun— this cute idea, that the universe is a big network, like the London Underground, that changes over time," Severini says.

Markopoulou was partly inspired by the principle of emergence, where complexity can emerge from simplicity, or, more to her point, simplicity from complexity, such as wiggling water molecules forming ice crystals or waves. Paramount in her model was the ability to create images that explained *geometrogenesis,* her and her colleagues' term for the emergence of the structure of space-time during a critical phase in the birth of the universe. "Once it starts being hard to visualize, I'm not happy, I get uncomfortable," she says. "I also think you can have an extreme richness while staying with very few building blocks."

She was tickled by an aperçu from Ludwig Boltzmann, the 19th-century Austrian physicist, who looked at the physical properties of atoms and said, "Every Tom, Dick, and Harry felt himself called upon to devise his own special combination of atoms and vortices, and fancied in having done so that he had pried out the ultimate secrets of the Creator." Markopoulou chuckles. "It felt to me, when we were arguing 'Is it my model? Is it your model?' we were totally every Tom, Dick, and Harry."

In quantum graphity, space evolves out of dots that are either "on" or "off"—connected or disconnected to the next dot. It doesn't matter exactly what the dots are; they represent coordinates in a network of relationships, the fundamental constituents of the universe. "The value of this is in trying to give, however primitive it might be, some language to talk about space not being there," she says. The idea, Markopoulou explains, comes from a branch of mathematics known as category theory, in which "what something is, is the sum of how it behaves, rather than how it is." At the highest possible energy, at the beginning of the universe, all the dots in the graph are joined, and no notion of space exists. But as the system cools and loses energy, the points start detaching, which creates the dimensions and laws of space. In this model, space becomes like a crystal that forms out of a liquid as it cools.

"It was very courageous of Fotini to start working on this," says Sabine Hossenfelder, a research fellow at the Frankfurt Institute for Advanced Studies, a think tank devoted to theoretical physics, who from 2006 to 2009 did postdoc work at Perimeter. "It's the kind of thing you think has been done long ago, but surprisingly it wasn't." It would have been much easier, Hossenfelder says, for Markopoulou to find a niche for herself within an existing theory, like loop quantum gravity. "But quantum graphity of course is much more exciting. It's a new idea, one that could have done a good job bridging the gap between theory and experiment."

As Markopoulou's reputation grew, she was often called upon to represent Perimeter to the public. She was a young, accomplished woman in physics—a rarity. She enjoyed and tended to accept speaking invitations, partly to help change perceptions of female scientists. "For previous generations, the question was 'Are there women in science?' Now there are, but girls want to know, 'Are they normal?' When you seem to be happy, and you seem to be a woman they'd be happy to be, that's a fairly big thing." Her world revolved around quantum gravity. Shortly after separating from Smolin, she had fallen into a relationship with a German postdoctoral fellow at the institute, Olaf Dreyer, whom she married after four years. They lived and breathed their discipline. "It's nice to share these things with somebody closely," she says.

But Markopoulou found her more radical theories were sometimes greeted with the sly criticism that they were "creative." "The fact that you don't look like the standard makes it hard for them; they will take longer to form judgments, which means you stay in the doubt area for longer," she says. It was made worse by the pervasive attitude among physicists that you should gather your laurels by doing sensible calculations throughout your career and only cook up new theories of quantum gravity in your old age.

Markopoulou refused to play that game, and a sense of discontent began to build. Every problem she solved created a quagmire of fresh ones; and, daughter of sculptors, she was tiring of academic papers as the only tangible thing she could "make." After a while even her public-facing activities began to grate. "There is a part of me that felt like a kind of clown, telling people magical things about the universe," she says. "Something you take very seriously and you've devoted your life to, and you've made your own sacrifices for, is for other people, at best, entertainment." She consoled herself with something Isham once told her, counsel he'd received in turn from the physicist John Archibald Wheeler. When it comes to quantum gravity, he says, you're bound to fail. "What's important is not the fact you fail, but how you fail." Markopoulou was determined to fail better, to borrow Samuel Beckett's phrase.

In 2008 the South African physicist Turok was appointed director of Perimeter. Turok, who describes himself as "very demanding," pulled back from the more outré flavors of foundational physics and expanded into other areas, including particle physics, cosmology, and quantum computing. He didn't want the institute, he says, "to be a center for alternative physicists who were doing unusual things in speculative directions."

By 2009, Markopoulou's personal life was undergoing its own quantum transitions. At a conference in Waterloo about physics and the financial crisis, organized by Smolin, Markopoulou met systems theorist and physicist Doyne Farmer. The pair was instantly dazzled by one another, and within five days decided to upend their lives to be together. Markopoulou separated from Dreyer and Farmer from his wife. Not long after, she got pregnant on a road trip from San Francisco to Santa Fe in Farmer's 1967 Datsun convertible. Their son, Maris, was born in 2010; a year later Markopoulou's mother, Maria, passed away.

Markopoulou was still attracted to deep inquiry, but the further down she went, the less objective she found her colleagues' judgments about the value of her work. "If you're in a place where everything is certain, that's a very boring place," she says. "But if you jump out with no parachute, it's either a sociological exercise or a folly." She'd been striving to position her research at the metaphorical "edge of chaos," the point at which order emerges from complexity. But she'd started to get the creeping suspicion she was back with the artists in her mother's studio, competing for recognition and influence without any clear standards.

"In the absence of any kind of experimental confirmation or the ability to falsify your theories, quantum gravity has ended up being dominated by a few influential tastemakers," says Lloyd. "Fotini fell foul of that because she had her own strong sense of what is a good thing to do; her tastes were different."

As the institute continued to grow, Turok faced the challenges of needing to formalize its processes and manage larger numbers of physicists. Perimeter had already begun the process of implementing tenure for its faculty, which Turok inherited. Markopoulou prepared to apply. By this time she was back in Berlin again, on a fellowship at the Max Planck Institute. She put together a dossier of her accomplishments for Turok, which was also to be reviewed by a tenure committee and quantum gravity experts.

Turok says he respected Markopoulou but doubted her work would lead anywhere. He denied her tenure. "Fotini had pursued a very independent line of inquiry that was really very different and hardly acknowledged by leading researchers in the field," Turok says. "I applauded her for her bravery for pursuing her own line, but that inevitably brings risks with it. She is a very fundamental thinker; she had original ideas. But at the end of the day you had to decide if those ideas are going to pan out."

Markopoulou says she was disappointed that Perimeter had "shifted from a flat hierarchy of scientists to an all-powerful director." Turok emailed her out of the blue, she says, to stop the review process and deny her tenure. "As a result of my case, an independent consultant was appointed, because I had been the only woman faculty for nine years, I had a strong academic record, and Neil stopped the tenure process just as I had a baby," Markopoulou says. ("I respectfully beg to differ," Turok responds. "A tenure review process was never initiated.") The matter is subject to an out-of-court settlement.

In the autumn of 2011, Markopoulou walked out of Perimeter for good.

One sunny morning in March, I visit Markopoulou at her home outside Oxford, perched on a hill and encircled by stands of oak, ash, and silvery birch. Nancy, the wife of the poet and classicist Robert Graves, used to run a grocery store on the site before the poet John Masefield knocked it down to build a theater in 1924. The top rooms sit snug under the original proscenium arch. Markopoulou loves theater—a legacy of being Greek, she says. It allows you to "step out of your normal shoes, to shift reality a bit, and to actively participate by forcing you to suspend your belief." Not unlike science at its highest levels.

In the living room, Markopoulou bundles herself up into a burgundy armchair with the cheerful self-possession of a family cat. A Persian rug sprawls across the wood floor, monopolized by a Lego space station. One of her mother's bronze busts broods from a windowsill, a beautiful, dauby figure of a woman with braids parted down the middle. "You were asking me the other day what made me change," she says. "One big thing was my mother died and opened up a space for me." Markopoulou wouldn't have touched art while her mother was alive, but recognizes now that a similar desire to make, to craft and to create, is part of who she is.

"In many ways, physics and what I did are almost ideally positioned to my experience with my mother," she says. "I probably did come out of that wanting a much more firm grasp of what is what and objective decision-making. Now I don't feel I need it that much anymore, but growing up that was a big deal. Also, it was far away from her, it was my own space, but at the same time there were many ways in which the deeper challenges are the same." Sculpture is a lot like creating a physics theory, she says, because you have to turn it around and make sure it works from every angle. "You have to understand the essence of what you're doing before you start, because only then do you have a chance that it's going to work from all sides."

Did she believe this philosophy could help her solve quantum gravity? "I never really wanted to single-handedly solve it. But I

never went in thinking that we can't. I always assumed it was possible." Does she still think it is? "Not soon, but I don't know. If I
knew, I would be doing it," she quips.

During her unsteady transition out of physics, Farmer was a pil-
lar of support for Markopoulou. "I'm very much what I do, so going through a transition is a time when I don't know who I am,"
she says. "I was lucky to have the context where that was perfectly
possible."

Farmer is a distinguished and idiosyncratic physicist in his own
right. While still in grad school at the University of California,
Santa Cruz, studying physical cosmology, he and fellow physicist
Norman Packard created one of the world's first wearable computers. Released in the 1970s, it was a toe-operated device embedded
in the tip of a shoe, which allowed the wearer and an accomplice
to track the progress of a roulette ball and achieve a 20 percent
advantage over the house. He and Packard later decided to found
one of the first predictive stock-trading companies, which was ultimately sold to UBS, a financial services company, in 2006. Farmer's interests now lie in "econophysics," a field which he founded
and which applies the mathematics of natural systems to gather
insights about the economy.

When Farmer landed a post at Oxford University in 2011,
Markopoulou was faced with "the usual two-body problem in academia" of trying to find a job nearby. But when she started looking, she realized her heart wasn't in it. She had been toying with
industrial design, and asked the advice of the musician Brian Eno,
a physics follower and a friend. He advised her to look into the
master's program in innovation design engineering run by the
Royal College of Arts and Imperial College in London, and wrote
her a letter of reference.

She sailed through the admissions process, which included
an exercise where prospective students had to explain how they
would evade a pack of zombies chasing them toward the lip of a
cliff. She enjoyed the classwork but found the mental shift hard at
first. "It just felt silly, because you go from 'This is how the universe
started' to 'This mattress has these bubbles.'"

But she loved making things, and also made the personal connections that evolved into Doppel. A sailing trip to Greece, in
which the Doppel crew nearly scuppered Markopoulou's and
Farmer's large-bottomed boat on the rocks off the island of Ceph-

alonia, cemented the team's conviction that they could withstand the trials of doing a startup.

The kernel for Doppel came from a piece in *New Scientist* about interoception—the way humans can discern the internal state of the body and conceive of it as "their own." The idea is that our sense of self is not merely a mental process that somehow envelops the body but somehow arises from the two-way conversation between the brain and other organs. As Manos Tsakiris, the softly spoken psychologist and neuroscientist who advises Doppel, tells me, "You cannot cut off cognition from the rest of the body, and you cannot cut off the body from the rest of the world in which you interact." By harvesting your natural response to rhythm, Doppel runs counter to the notion that the self resides in the mind alone—that the human is a creature of the will, a maker of rational decisions, a sovereign mind bossing around dumb matter.

With hindsight, Markopoulou sees her work at Doppel as a "natural evolution" from what she did before. Isham had inspired her to pursue physics as a quest to understand reality from within, when scientists can't stand apart from what they're trying to analyze. But now, instead of the universe as the ultimate system, she has the human body. "If you think about physics, it's a human creation. The equations represent stuff we come up with because of our senses. So shouldn't our senses be part of what goes into physics?" Markopoulou says.

Markopoulou thinks that many disputes in science come down to competing metaphysical commitments. She recognizes that her own belief in the fundamental nature of time, and her dislike of timelessness, are moral preferences as much as anything else. "Most of the physics where time does not exist comes with determinism as well. There is something about thinking that time is real and being responsible for your actions," she tells me.

This belief in the inexorable movement of time is what seems to have allowed Markopoulou to reinvent herself—to turn away from years spent building a career as a physicist and to start from scratch as a designer and entrepreneur. "This is the nice thing about me, but it's also a little bit weird: when I do something, I just do it. So when I switched, I switched," she says. "Our mind can live in the past, the future, or any fantasy place it wants, but our body only processes the now."

Doppel is unlikely to be the end of Markopoulou's journey.

"Whatever it is that you do, it has to have a context. Academia is one context, business is another context. I can't really tell you if it's better or worse, it's a different set of rules—and right now I have come to no conclusions as to what I think about those rules. I'm still exploring."

DAVID EPSTEIN

The DIY Scientist, the Olympian, and the Mutated Gene

FROM *ProPublica*

TWO YEARS AGO I wrote a book called *The Sports Gene* that examines the intersection of genetics and athleticism. I expected my mother to buy a dozen copies and invite me to her book club and that would be the end of it. (She did.) Instead I was almost immediately bombarded with emails from people wanting to know if their kid has Serena Williams's genes. One coach emailed, wondering how one would get athletes involved in genetic experimentation.

They were coming so quickly, and many were so unhinged, that I took a brief break from opening them.

And then I got one that had this subject heading: "Olympic medalist and muscular dystrophy patient with the same mutation." Now that caught my attention. I wondered if it might point me to some article or paper in a genetics journal about an elite athlete I'd somehow missed.

Instead it was a personal note from a 39-year-old Iowa mother named Jill Viles. She was the muscular dystrophy patient, and she had an elaborate theory linking the gene mutation that made her muscles wither to an Olympic sprinter named Priscilla Lopes-Schliep. She offered to send me more info if I was interested. Sure, I told her, send more.

A few days later I got a package from Jill, and it was . . . how to put it? . . . quite a bit more elaborate than I had anticipated. It included a stack of family photos—the originals, not copies;

a detailed medical history; scientific papers; and a 19-page illustrated and bound packet. I flipped through the packet, and at first
it seemed a little strange. Not ransom-note strange, but there were
hand-drawn diagrams with cutouts of little cartoon weightlifters
representing protein molecules. Jill had clearly put a lot of effort
into this, so I felt like I had to at least read it. Within a few minutes
I was astounded. This woman knew some serious science. She off-
handedly noted that certain hormones, like insulin, were too large
to enter our cells directly; she referred to gene mutations by their
specific DNA addresses, the way a scientist would.

And then I came to page 14.

There were two photos, side by side. One was of Jill, in a royal
blue bikini, sitting at the beach. Her torso looks completely normal. But her arms are spindles. They almost couldn't be skinnier,
like the sticks jabbed into a snowman for arms. And her legs are so
thin that her knee joint is as wide as her thigh. "Those legs can't
possibly hold her," I thought.

The other picture was of Priscilla Lopes-Schliep. Priscilla is one
of the best sprinters in Canadian history. At the 2008 Olympics in
Beijing she won the bronze medal in the 100-meter hurdles. It was
the first Canadian Olympic medal in track and field since 1996. In
2010, Priscilla was the best 100-meter hurdler in the world.

The photo of her beside Jill is remarkable. Priscilla is in mid-
stride. It's difficult to describe just how muscular she looks. She's
like the vision of a superhero that a third grader might draw.
Oblong muscles are bursting from her thighs. Ropey veins snake
along her biceps.

This is the woman Jill thought she shared a mutant gene with?
I think I laughed looking at the pictures side by side. Somehow,
from looking at pictures of Priscilla on the Internet, Jill saw something that she recognized in her own much smaller body and decided Priscilla shares her rare gene mutation. And since Priscilla
doesn't have muscular dystrophy, her body must have found some
way "to go around it," as Jill put it, and make enormous muscles.

If she was right, Jill thought, maybe scientists could study both
of them and figure out how to help people with muscles like Jill's
have muscles a little more toward the Priscilla end of the human
physique spectrum. Jill was sharing all this with me because she
wasn't sure how best to contact Priscilla and hoped I would facilitate an introduction.

It seemed absolutely crazy. The idea that an Iowa housewife, equipped with the cutting-edge medical tool known as Google Images, would make a medical discovery about a pro athlete who sees doctors and athletic trainers as part of her job?

I consulted Harvard geneticist Robert C. Green to get his thoughts, in part because he has done important work on how people react to receiving information about their genes. Green was open to discussing it, but he recalls a justifiable concern that had nothing to do with science: "Empowering a relationship between these two women could end badly," he says. "People go off the deep end when they are relating to celebrities they think they have a connection to." I was skeptical too. Maybe she was a nut job.

I had no idea yet that Jill, just by investigating her own family, had learned more about the manifestations of her disease than nearly anyone in the world, and that she could see things that no one else could.

Jill was born in 1974, and she met all the normal baby milestones — sitting, crawling, and walking right on time. She was tiny, though. Not short, just slight. She was so skinny as a three-year-old that she once fell into a toilet bowl, her feet sticking over her head. The stumbling didn't start until she was four.

Her preschool teacher noticed it first. Pretty soon it morphed into full-fledged face-planting. Little Jill told her mother she was afraid of witches' fingers. "When I walked I'd feel the sensation of there were almost little gnarled hands and fingers reaching up and grabbing my shins," Jill says, "and I'd fall really forcefully."

Jill's dad remembered having some trouble walking as a kid, and his doctors told him he'd had a very mild case of polio. But Jill's symptoms were much more pronounced, and her pediatrician was stumped. He told the family to go to the Mayo Clinic.

They were stumped there too. They tested the entire family and saw that Jill, her father, and her brother had higher-than-normal levels of creatine kinase in their blood. That is an enzyme that leaks out of muscles when they are damaged, but Jill was the only one struggling to walk. Based on the creatine kinase, the doctors thought the family might have some form of muscular dystrophy, but that didn't usually show up this way in little girls, and why did Jill's father and brother seem fine?

"They were really sure they'd never seen anything like this," Jill says. "They said our family was extremely unique, and they couldn't define what type it was. And ultimately that's good in one way, because they're being honest. But on the other hand, it was terrifying . . . It's alarming if you don't have something to grasp a hold of."

Jill returned to the Mayo Clinic every summer, and it was always the same. There was nothing doctors could do and nothing new they could tell her. The constant falling on her face stopped on its own, but it was replaced by a burning sensation in her legs. And while Jill was growing in height like a normal girl, the fat on her arms and legs was vanishing. By the time she was eight, her arms and legs were so skinny that other kids would wrap their fingers around them and ask if her mother fed her at home.

By 12, veins were starting to pop out of her legs, and the other kids started asking how it felt to be old. She was rail-thin, but she could still do most of the things normal kids did. A video of her 12th birthday shows Jill at a pool party, her cannonball displacing a teacup of water.

Later that year her muscles started to fail again. "I can remember just getting on a bike I'd always ridden," Jill says, "and feeling like someone came up behind me and just threw me into the handlebars." Suddenly she couldn't hold her upper body up over the bike. She invited a friend to go roller skating and found that she couldn't stand up on the skates. Over the course of a few weeks, Jill had completely lost the ability to ride a bike and skate.

Something was "terribly wrong," as she put it, but she didn't even bother to tell her parents about it. Other people went to doctors and got solutions. That had never happened for Jill, so she started looking for answers on her own, the way a kid would. She started bringing home books from the library on poltergeists and other supernatural phenomena. "I remember it really freaked out my dad at one point," she says. "He was like, 'Well, are you into the occult, or what?' It was nothing of the sort." It was just that she couldn't explain the forces acting on her body. She was fascinated by the stories of people bedeviled by inexplicable maladies or situations. Jill says, "Ya know, I believe them."

By the time she left for college, Jill had maxed out at five-foot-three and 87 pounds. She had long since moved on from polter-

geists, but not from the knowledge that if she was going to figure
out what was happening to her body, she would have to go it alone.

Almost as soon as she arrived on campus, she hit the library. She
spent more time there than in classes, about 25 hours a week as
she recalls it. Twenty-five hours a week just poring over every textbook and scientific journal she could find on muscle disease. She
did this for months, going article by article, like a police officer
driving up and down every street doing a grid search. But nothing
quite fit. Not until she came to a paper in the journal *Muscle and
Nerve* on a rare type of muscular dystrophy called Emery-Dreifuss.
"Looking at the pictures," Jill says, "it was a very startling thing to
realize I'm seeing my dad's arm."

Jill's dad was thin, but the muscles in his forearm and hand
were unusually well defined. Jill would call it a "Popeye arm" when
she was a little girl. In another paper she saw that Emery-Dreifuss
patients often had that same trait; it was even referred to as a Popeye arm deformity. But she didn't see pictures of women with the
disease.

The *Muscle and Nerve* paper described the three hallmarks of
Emery-Dreifuss patients: they couldn't touch their chins to their
chests or their heels to the floor, and their arms were perpetu-
ally bent at the elbow. The medical term is *contractures*. In middle
school, Jill's head once got stuck for a while so that she was looking up.

"I'm getting chills reading this," Jill says. "I've got all three." Picture a Barbie doll—arms always bent, feet slanted to fit into high
heels, and a stiff neck. It's "very ironic that we give this to little girls
and say this is perfect, and we're actually handing them a doll that
has a genetic disorder."

Jill was positive that this is what she had. So she read on, and got
scared. The papers noted that Emery-Dreifuss always comes with
heart trouble. She had to read more, so before going home for
college break she stuffed her bag full of medical books and papers.

She didn't want to scare anyone in the family, so she didn't
read the papers openly. But one day during break she went to the
kitchen to microwave popcorn and returned to find her father pe-
rusing the stack of material from her bag. He told her he had all
the symptoms she was reading about. "Well, yeah, I know . . . the
arm, and the neck," Jill told him. No, he replied, all the cardiac
symptoms.

Doctors had told Jill's father years earlier that his irregular heart rhythms had been due to some kind of virus. "It's not," Jill told him. "We have Emery-Dreifuss." She went to the Iowa Heart Center with her Emery-Dreifuss papers and started insisting that a cardiologist needed to see her father. At first nurses told her she'd need a referral. But Jill had trouble getting one and was so relentless that eventually they gave in. The cardiologist put a Holter monitor on Jill's dad, which tracked his heart's electrical activity for a day. At one point his pulse rate dropped into the 20s, which meant he was either about to win the Tour de France or he might be about to drop dead. He was 45, and had a pacemaker put in immediately.

"She saved her dad's life," Jill's mother, Mary, says. "If it wasn't for her, how would we have ever known this? . . . But I think it was a hard burden for her, because it seemed like no one else was looking."

Even after the pacemaker surgery, the Iowa Heart Center couldn't confirm that Emery-Dreifuss ran in Jill's family. She had, though, in her reading, come across a group of researchers in Italy who were looking for families with Emery-Dreifuss to study, hoping to locate a gene mutation that causes the disease.

So 19-year-old Jill put on her most serious navy pantsuit, again gathered up her papers, and took them to a neurologist in Des Moines. She asked the neurologist to take a look, hoping that she would help her connect with the Italian team and get in the study. But the neurologist would have none of it. "No, you don't have that," Jill recalls the neurologist saying sternly. And then she refused even to look at the papers. It might seem rude that a doctor refused just to hear Jill out and glance at the papers, but at the time most doctors believed that Emery-Dreifuss only occurred in men. Plus this was a self-diagnosis of an obscure disease coming from a teenager.

So Jill wrote to the Italians herself. She constructed a family tree, noting all the symptoms she saw in her father, two younger brothers, and a younger sister, and then she stripped down to her underwear. "I set the timer on my camera and I took pictures of myself," Jill says, "because I thought, Well, if that's how I identified it, let me send a picture."

Up to that point the Italians had only collected four other families to study, so they were thrilled to hear from Jill and immediately wrote back. From the letter it seems as though the Italian team thought Jill had access to a lab. Can you send DNA from your entire family? it reads. "If you cannot prepare DNA, just send fresh blood." And then it gave mailing instructions.

Discouraged by her encounter with the neurologist, Jill figured it would be a dead end to show up at a hospital and ask that her blood be drawn so she could ship it to Italy. So she convinced a nurse friend to smuggle needles and test tubes to her house. They filled them with her family's blood. At the post office, when Jill declared that her packages contained blood, an employee had to retrieve a big binder that listed what can be shipped to various countries. Fortunately, Italy took blood in the mail.

Today an entire human genome can be sequenced in a few days. But in the mid-1990s sequencing was a ponderous ordeal. It would be four years before Jill heard back from the Italians.

In the years after she sent her family's blood overseas, Jill waited for confirmation that she had Emery-Dreifuss. Sort of.

She was so confident that in her annual trips to the Mayo Clinic she started taking a pen from her purse and writing "Emery-Drei-fuss" on her medical chart. Her mom would get upset: "You cannot change your chart!" I want what I actually have to be listed, Jill would tell her.

Then in 1999 Jill got an email from Italy. She stopped before opening it to let the moment sink in. And then she clicked. She had a mutation on a gene known as LMNA, or, for ease, the lamin gene. So did her father, two brothers, and a sister. So did the other four families in the study with Emery-Dreifuss.

Jill had been right about her self-diagnosis, and the researchers discovered the responsible gene mutation. What makes the lamin gene so important is that it carries a recipe for constructing the nuclear lamina, a tangled net of proteins at the center of every cell —one that influences how other genes are flipped on or off, like light switches, changing how the body builds fat and muscle. Mutations in the lamin gene are known to cause more than a dozen diseases, from a gradual loss of physical sensation to a hyperrapid aging known colloquially as "Benjamin Button disease."

Jill's lamin gene has a typo in it, a serious one. We all have little genome typos, or mutations, hundreds of them. *Mutation* is simply the term for a version of a gene that fewer than 2 percent of the population has. But most of them don't do anything. Along the spiraling ladders of her DNA, the 3 billion G's, T's, A's, and C's in Jill's genome, the single-letter typo on her lamin gene just happened to be in a very unfortunate place. To appreciate the scale of what happened to Jill, imagine enough letters to fill 13 complete sets of *Encyclopaedia Britannica* with a single-letter typo that changes the meaning of a crucial entry.

Jill was glad that she had been right about Emery-Dreifuss and that she knew her mutation. But still, "it's almost darkly comical," Jill says. "It comes down to a G that was changed to a C."

Soon after she got her genetic results from Italy, the research group published a paper in the journal *Nature Genetics* showing that a mutation on the lamin gene can cause Emery-Dreifuss. They thanked Jill at the bottom of the paper. (Her last name at the time was Dopf.)

Jill was 25, and a lab director at Johns Hopkins University had heard through the medical grapevine about the young woman who diagnosed her own Emery-Dreifuss. Wanting both a dogged intern and—why not?—a real-life lamin mutant in her lab, the scientist offered Jill a summer internship. Jill's job was to sift through scientific journals and find any references to diseases that might be caused by a lamin mutation.

Sitting there day after day reading, as she had freshman year in college in the library, Jill came across an incredibly rare disease. A disorder called partial lipodystrophy. It caused fat on certain parts of the body, particularly the limbs, to disappear, leaving veins and muscles to stand out as if they'd been shrink-wrapped in skin. Looking at photos of patients with partial lipodystrophy, all Jill could think was that they looked like her family members.

Could Jill have not just one but two incredibly rare genetic diseases? The odds of Jill's having Emery-Dreifuss were so rare that the prevalence isn't even known; certainly more rare than one in a million. The odds of having partial lipodystrophy are probably somewhere between one in one million and one in 15 million. The odds of separately getting both by chance alone? It was one in far more than the number of people who have lived on Earth, ever.

Jill attended a medical conference at Hopkins during her internship, and, as she had with Emery-Dreifuss, she showed photos to doctors and told them she thought she had partial lipodystrophy. Just like before, they assured her it wasn't the case. They jokingly diagnosed her with something a lot more common: intern syndrome. "Where you have a medical student being introduced to a lot of new diseases," Jill says, "and they keep thinking they have what they're reading about."

This time Jill believed the experts, so she dropped it. One rare disease was enough. She went back to reading about Emery-Dreifuss. But pretty soon she dropped that too. She was learning more about all the cardiac problems—the average lifespan of subjects in case studies she read was around 40—and the stress landed her in the hospital. "I had two panic attacks that were brought on by the stress of reading all these things," Jill says. "And I went to a counselor for a while and worked with my cardiologist. And we decided it was just too much information. It wasn't healthy."

So she stopped reading scientific literature. Cold turkey. No more medical research. No more DIY diagnosis. She started working as a writing instructor at community colleges and taught adult education at night.

She started dating and met Jeremy, the man she would marry. And though there was a 50-50 chance she would pass down her Emery-Dreifuss gene mutation, they decided to have a child. Jill's pregnancy was normal, and her son, Martin, did not inherit the mutation. But after he was born, Jill's physical problems accelerated.

She started having muscle twitches—not just little ones, in one muscle at a time, but from head to toe, for hours. Suddenly she was having to hold on to Jeremy to steady herself when she walked. "The best way I can describe it," Jill says, "would be like where gravity is getting incredibly heavy."

By Martin's first birthday she could hardly walk. One day he was calling that he wanted mac and cheese. It was just a few feet to the kitchen. "I had six steps to take," Jill says, "and I realized this is it. This is the last six steps I'm going to take." After that Jill could not get up again.

Her father was losing his ability to walk at exactly the same time, so father and daughter transitioned to life in motorized scooters.

Jill remembers seeing her father discouraged for the first time in his life. After one visit with a neurologist he told her, "I feel like I go there just to be weighed."

Five years later Jill's father told Jill's mother that he was tired, so he moved from his scooter to his favorite chair. He bowed his head as if taking a nap, but he never woke up. His heart had finally failed, at the age of 63.

On the day her dad passed away, Jill and her siblings and a few relatives had dinner together at her mom's house. Later that week Jill's younger sister Betsy pulled Jill over to the computer to show her a picture.

People often asked Betsy what kind of workout she did, because the muscles in her arms were so well defined. But it wasn't from the gym. Betsy's arms had always been defined, and as she grew up, she wanted to know why. Jill told her that she might want to look into lipodystrophy but that doctors had told her years ago she didn't have it. Betsy attended a meeting for people with lipodystrophy and there learned about an Olympic sprinter who was conspicuously missing fat. The picture Betsy showed Jill was, of course, of Priscilla Lopes-Schliep.

"She pulls up pictures of this extremely muscular athlete," Jill says. "And I just took one look at it, and just . . . what?! We don't have that. What are you talking about?"

But a week later, with the funeral out of the way, Jill got curious. She started Googling. Not just pictures of Priscilla running, but photos of her at home, just hanging around or feeding her baby daughter. She saw the same prominent veins, the same fall of clothing over shoulders and arms missing fat. The same visible divisions between muscles in the hips and butt. "It was just unmistakable," Jill says. "It's like a computer that can analyze a photo and get a match and be 100 percent sure that's the same shoulder, that's the same upper arm. I see the same veins, I see them branching this way. You just know and it's hard to convey, how could you just know. But I knew we were cut from the same cloth. A very rare cloth."

It was the third time Jill had made a visual lock on something rare. First it was with her family's Emery-Dreifuss, then when she thought they had lipodystrophy, and now she thought that she and Priscilla just must have a mutant gene in common because of the exact same pattern of missing fat. But how then did Priscilla get a double helping of muscle while Jill's muscles were scarcely there?

"This is my kryptonite, but this is her rocket fuel," Jill says. "We're like comic-book superheros that are just as divergent as can be. I mean, her body has found a way around it somehow."

If this sounds familiar, it might be because it's basically the plot of the Bruce Willis and Samuel L. Jackson movie *Unbreakable*, about comic-book superheros. Jackson plays a broken-boned, physically fragile man searching for his genetic opposite, a man born so strong he can survive any physical trauma. In the final scene, Jackson tells Willis he's the guy.

Jill's 12-year abstinence from medical literature was over; she wanted to enlist Priscilla in her genetic detective work. Partly out of curiosity, but also because if Priscilla's body had indeed found some way to "go around" a lamin mutation that should cause muscle failure, it could be important for scientific research.

There was just one practical problem. Jill had no idea how to go about reaching Priscilla so that she could tell her all this. "I've had crazy ideas, like can I show up to Canada at a meet-and-greet or a track event?" Jill says. "You're just going to get a restraining order at best. I mean, people will think you're crazy if you're on this motorized scooter and you're going up to this hurdler. Nobody in security is going to let you by."

A full year passed. That's when I came in.

Jill happened to be in earshot of her television when I started yammering about athletes and genetics on *Good Morning America*.

"I thought, 'Oh, this is divine providence. This is exactly what I'm looking for,'" Jill says. So she sent me an email, and then the package with original family photos, scientific papers, and the 19-page bound packet explaining her gene mutation and her theory that Priscilla also had a lamin mutation. Jill wanted my help getting in touch with Priscilla.

It just so happened that Priscilla's agent and I followed one another on Twitter. So I sent him a direct message. I didn't expect anything to come of it. I mean, I was telling a pro athlete that a stranger in Iowa wanted to talk to her about getting a genetic test. Luckily, Kris Mychasiw, Priscilla's agent, is an extraordinarily nice guy. And he also knew that Priscilla constantly faced steroid whispers because of her musculature. After she won the Olympic bronze in 2008, some media in Europe began to openly accuse her. She ran at one meet in France specifically so that she could address the French media.

Kris passed on my request to Priscilla. "He was just like, 'This lady in Iowa. She says she has the same gene as you, and wants to have a conversation,'" Priscilla recalls. "I was kind of like, 'Um, I don't know, Kris.'" But he told her just to have the conversation and see where it goes.

I talked to Priscilla first, and then she and Jill talked on the phone. Jill sent Priscilla the same 19-page packet she'd sent me. And it wasn't any of the abstruse science that caught Priscilla, it was the childhood stories Jill shared about kids pointing at the veins in her legs. As a little girl, Priscilla would come home asking her parents to get the veins removed from her legs because the boys were making fun of her. One of Priscilla's cousins had looked into lipodystrophy, but nobody in the family had ever gotten a firm diagnosis, nor did they know much about the condition. They just knew there were a lot of strong and well-defined people in the family, especially the women. But there wasn't much impetus to do a deep investigation. After all, when Priscilla was a kid she wasn't falling like Jill, she was getting strong, and fast. She earned a track scholarship to Nebraska, where she became one of the best athletes in the university's history and won a national championship. After college she turned professional and won her Olympic medal.

Still, when Priscilla would walk around at track meets, she'd hear people commenting, she says, "Oh look at her glutes, look at her arms, shoulders, calves. Oh look, look, look!" A picture of a male bodybuilder's face was pasted onto a photo of Priscilla's body —while she was straining to the finish line, attempting to make the Olympic final—and posted online. "That was pretty messed up," Priscilla says. "I was really pissed off about that . . . A lot of people honestly believed I was taking steroids."

Priscilla thinks that because of her physique, she was targeted for more than the normal amount of drug testing. (Targeted testing is a standard part of antidoping.) She was tested right after having her daughter, Natalia. At the World Championships in Berlin in 2009, she was tested just minutes before winning a silver medal. There's not even supposed to *be* any drug testing that close to the race.

The following month, at a meet in Greece, someone stole her training journal out of her bag. It was at the very bottom, underneath expensive workout clothes and shoes, none of which were taken. Why steal a training journal? We'll never know. But I've covered a lot of doping stories, and I'm convinced someone thought the journal contained her steroid regimen.

Jill and Priscilla spoke on the phone several times. Then, eight months after I introduced them, they agreed to meet in person. They picked a hotel lobby in Toronto, where Priscilla lived. Jill arrived first, with her mom.

She watched the clock. It ticked past the time they were supposed to meet. Jill got scared. This was a crazy thing she was doing. What if Priscilla had decided not to show? She watched the door. And when Priscilla walked in, Jill's first thought was, "Oh my gosh, it's like seeing family."

Priscilla felt the same way. "It really was just a wow moment," Priscilla says. "Like, do I know you?" The two women started flexing for one another, Priscilla's muscles many times larger but with the same definition exposed by a lack of fat. They even retreated to a hallway in the hotel to compare body parts. "There is something real here," Priscilla recalls thinking. "Let's research. Let's find out. Because how could the gene do this to you and this to me? That was what my question was. How?"

Jill offered Priscilla a cashier's check, money that had been raised for research in a memorial fund after her father's death. Jill hoped Priscilla would take it and use it to pay for a genetic test. And Priscilla agreed.

It took a year to find a doctor to test Priscilla. She visited several clinics. Some told her they just didn't do that test. Others said they weren't sure how to interpret the results, so they felt it wouldn't be responsible to do the test.

Finally Jill went to a medical conference and approached the foremost expert in lipodystrophy, Dr. Abhimanyu Garg, who runs a lab at the University of Texas Southwestern Medical Center. He agreed to do both genetic testing and a lipodystrophy evaluation.

The results showed that Jill had been right. She and Priscilla do have a genetic connection. Not only do she and Priscilla both have lipodystrophy—the disease Jill had been told to cast aside back when she was an intern at Johns Hopkins—but they have the

exact same subcategory of partial lipodystrophy, known as Dunnigan-type.

And Priscilla did indeed have a mutation on her lamin gene.
Both women have a typo on the same one of their 23,000 genes.
Priscilla's is not the exact same "single-letter" typo that Jill has,
though; it's a neighbor typo. That splinter of distance in typo location seems to makes the difference. It's why Jill has Emery-Dreifuss
and Priscilla has fantastic musculature. (That said, there are people with Priscilla's exact genetic typo who have both fat and muscle
wasting.)

Dr. Garg called Priscilla immediately to give her the news. He
caught her at the mall, shopping with her kids. "I was just dreaming about going out and getting a juicy burger and fries," Priscilla
says, "and Dr. Garg calls me and says, 'I have your results.'" Priscilla
asked if she could call him back later, after lunch. He said that she
could not. "He's like, 'You're only allowed to have salad. You're on
track for a [pancreatitis] attack.' I was like, 'Say what?'"

Despite her monstrous training regimen, Garg informed Priscilla that due to her unmonitored lipodystrophy, she had three
times the normal level of triglycerides, or fat, in her blood. Garg
has had a number of partial lipodystrophy patients with unusually
large muscles who were good athletes, but none as muscular nor
as athletically accomplished as Priscilla. He thought her training
would probably protect her from the buildup of fat in her blood,
but he was wrong. "I thought with her physical training, I'm not
going to find much metabolic abnormalities," Garg says. "But this
was the one thing that was a severe problem on her blood testing."

In other words, Jill had once again helped steer someone away
from a medical disaster. She had prolonged her dad's life, and
now—once again with that cutting-edge medical tool Google Images—she caused the most intense medical intervention that a
professional athlete had ever had. Priscilla called Jill to tell her. "I
was like, 'You pretty much just saved me from having to go to the
hospital!'" Priscilla says. "Dr. Garg told me I have the gene and my
numbers are out of the roof."

Even Garg was startled by what Jill had done. "I can understand
a patient can learn more about their disease," he says. "But to
reach out to someone else and figure out their problem also. It is
a remarkable feat there."

Priscilla has now overhauled her diet and started medication.

She was recently recruited to try bobsled for Canada, which—if all goes well—could give her a chance to be the sixth athlete ever to win a medal in both the Summer and Winter Olympics. And with her genetic test, she hopes the steroid whispers will die down too.

A little more than a decade ago, when the first full human genome was sequenced, some medical futurists and optimistic doctors prophesied that we'd all bring our genetic information to the doctor's office to get treatment personalized to our DNA. Not only has that not materialized, but for the most part, studies that scan entire genomes for disease-causing DNA "have not resulted in anything clinically useful," says Heidi Rehm, a geneticist at the Harvard Medical School.

It's not that there aren't plenty of genes that matter, but it turns out that unraveling genetic causes of disease is a whole lot more complicated than that wishful thinking of a decade ago. Rather than single genes causing a disease—or any trait, really—it's usually many small genes, each with a tiny effect, combining to influence a condition, and doing so in concert with lifestyle and other environmental factors.

Many rare diseases, though, actually *are* caused by single genes that alone have a large impact on biology. So those mutant genes have been far easier to locate. And sometimes when scientists can figure out what's causing a serious and rare disease, they can begin to untangle more common ailments. "As you begin to understand these pathways," Rehm says, "there will be milder versions of those rare diseases you can help to affect by understanding the outliers."

For example, research on a rare gene mutation which gave people such low cholesterol levels it was a wonder they were alive led to a treatment for high cholesterol. An Alzheimer's treatment may one day come from ongoing research on a small group of people in Iceland who have a version of a gene that protects their brains in old age.

Recently Rehm and a group of scientists started something called the Matchmaker Exchange; it's a kind of OkCupid for rare diseases, where people with uncommon conditions can be matched with other people with similar diseases and gene mutations, in the hope that it will spark new discoveries. After all, under normal circumstances Jill and Priscilla never would have ended up in the same doctor's office. A person with a rare disease in their family will often have seen more cases and different manifestations of the disease than any doctor has.

Rehm herself discovered that Norrie disease, which causes loss of sight and hearing, is not only a neurological but also a blood-vessel disease when she found a Yahoo message board where Norrie patients were all discussing their erectile dysfunction. "I think there is a cultural change," she says. "Physicians are recognizing the very important role of the patient in being not only an advocate for themselves but really a source of relevant information."

Dr. Garg, who has studied lipodystrophy for 30 years, says that Jill and Priscilla are the most extreme cases of muscle development he has ever seen in lipodystrophy patients—on opposite ends of the spectrum, of course. What might be causing that?

Jill and Priscilla don't have the exact same typo, or "point mutation." Because of that they have one condition precisely in common—Dunnigan-type partial lipodystrophy—and another that is divergent as can be—their muscles. But the mechanism behind their difference is an important mystery. It might not surprise you by now that in search of an answer, Jill hit the scientific journals.

She alighted on the work of a French molecular biologist named Etienne Lefai. He does extremely technical work on a protein with the less-than-mellifluous name SREBP1. SREBP1 has long been known to manage fat storage. After a meal SREBP1 is helping each of your cells decide whether to use the fat that just arrived for fuel or store it for later.

Lefai's team found, in animals, that a buildup of SREBP1 in the cell can lead to either extreme muscle atrophy or extreme muscle growth. And that was something Jill was interested in. She sent Lefai a two-line email with a question about his work. He thought it was from a scientist or PhD student and responded.

Soon Jill told Lefai about her own history and suggested that it is possible that he discovered the actual biological mechanism that makes her and Priscilla so different—SREBP1 interacting with lamin.

"Okay, that triggers a kind of reflection from my side saying, 'That's a really good question. That's a really, really good question!'" Lefai says, in a thick French accent. "Because I had no idea of what I can do with genetic diseases before she contacted me. Now I have changed the path of my team."

Since Jill first contacted him, he has learned that lamin proteins—which the body creates using instructions from the lamin gene—can interact with SREBP1. Now Lefai is working to figure out whether a lamin gene mutation alters the ability of lamin proteins to regulate how SREBP1 works, causing simultaneous loss of muscle and fat. It's possible, though certainly not assured, that his work could ultimately lead to new treatments.

Given how technical his work is, I asked Lefai if he had ever had someone from outside the science community influence his research. "In my life, no," he says. "People from outside coming and giving me hope? New ideas? I have no other example of this kind of thing. You know, maybe happen once in a scientific life."

It is the dream of many rare-disease patients to have a scientist orient his research agenda around them.

The first time Jill and I spoke, she told me that she knew there would be no treatment breakthrough in her lifetime. (Although I'm not so sure.) But she doesn't want what she learned to be lost, and hopes that maybe she'll have made a small contribution to some therapy that's developed for some other generation. She told me recently that she has proved her point, and she's thrilled that she was able to help Priscilla improve her own health.

The two women have stayed in touch. They talk about their kids. Priscilla is quite sure her daughters got her mutation. She can feel the difference between other kids and her own when she lifts them. Her girls are dense, with solid muscles.

In my years of reporting on genetics and athleticism, I only know of two other cases where rare versions of single genes were associated with elite athletic performance, and the other two debuted in medical journals.

Still, Jill told me that she's officially retiring, for good this time, from DIY diagnosis. She gave me the same line that athletes so often wield when they hang 'em up: I want to spend more time with my family.

Her mother, Mary, says it would be great if Jill could move on and just focus on the rest of her life. But do you think she'll really retire? I asked. "Probably not," Mary told me.

Lefai laughed when I asked if he thought Jill would quit. "Of course she will continue," he says. "I don't believe that retirement.

In the last email she told me she was in contact with people in New Zealand."

It's been two and a half years since Jill happened to hear me on television and decided to reach out. Before Priscilla agreed to get a genetic test, I hadn't really thought of all this as a story I would one day write. I just thought Jill deserved a response. But I don't believe she's retiring either.

Recently Jill sent me an email. She'd picked up on a tidbit in a very technical scientific paper about potentially reversing muscular dystrophy. "I don't want to read too much into this," she wrote me. "But of course I'm curious."

ANN FINKBEINER

Inside the Breakthrough Starshot Mission to Alpha Centauri

FROM *Scientific American*

IN THE SPRING of 2016, I was at a reception with Freeman Dyson, the brilliant physicist and mathematician, then 92 and emeritus at the Institute for Advanced Study in Princeton, New Jersey. He never says what you expect him to, so I asked him, "What's new?" He smiled his ambiguous smile and answered, "Apparently we're going to Alpha Centauri." This star is one of our sun's nearest neighbors, and a Silicon Valley billionaire had recently announced that he was funding a project called Breakthrough Starshot to send some kind of spaceship there. "Is that a good idea?" I asked. Dyson's smile got wider: "No, it's silly." Then he added, "But the spacecraft is interesting."

The spacecraft is indeed interesting. Instead of the usual rocket, powered by chemical reactions and big enough to carry humans or heavy instruments, Starshot is a cloud of tiny multifunction chips called StarChips, each attached to a so-called light sail. The sail would be so insubstantial that when hit by a laser beam, called a light beamer, it would accelerate to 20 percent of the speed of light. At 4.37 light-years away, Alpha Centauri would take the fastest rocket 30,000 years to reach; a StarChip could get there in 20. On arrival the chips would not stop but rather tear past the star and any of its planets in a few minutes, transmitting pictures that will need 4.37 years to return home.

The "silly" part is that the point of the Starshot mission is not obviously science. The kinds of things astronomers want to know

about stars are not the kinds of things that can be learned from
a quick flyby—and no one knows whether Alpha Centauri even
has a planet, so Starshot could not even promise closeups of other
worlds. "We haven't given nearly as much thought to the science,"
says astrophysicist Ed Turner of Princeton University, who is on
the Starshot Advisory Committee. "We've almost taken for granted
that the science will be interesting." But in August 2016 the Starshot team got lucky: a completely unrelated consortium of European astronomers discovered a planet around the next star over,
Proxima Centauri, a tenth of a light-year closer to us than Alpha
Centauri. Suddenly Starshot became the only semifeasible way in
the foreseeable future to visit a planet orbiting another star. Even
so, Starshot sounds a little like the dreams of those fans of science fiction and interstellar travel who talk seriously and endlessly
about sending humans beyond the solar system with technologies
that would surely work, given enough technological miracles and
money.

Starshot, however, does not need miracles. Its technology,
though currently nonexistent, is based on established engineering
and violates no laws of physics. And the project has money behind
it. Yuri Milner, the entrepreneur who also funds other research
projects called Breakthrough Initiatives as well as yearly science
awards called Breakthrough Prizes, is kick-starting Starshot's initial
development with $100 million. Furthermore, Milner has enlisted
an advisory committee impressive enough to convince a skeptic
that Starshot might work, including world experts in lasers, sails,
chips, exoplanets, aeronautics, and managing large projects, plus
two Nobel Prize winners, the U.K.'s astronomer royal, eminent
academic astrophysicists, a cadre of smart, experienced engineers
—and Dyson, who, despite thinking Starshot's mission is silly, also
says the laser-driven sail concept makes sense and is worth pursuing. On the whole, few would make a long-range bet against an
operation with this much money and good advice and so many
smart engineers.

Whatever its prospects, the project is wholly unlike any space
mission that has come before. "Everything about Starshot is unusual," says Joan Johnson-Freese, a space policy expert at the U.S.
Naval War College. Its goals, funding mode, and management
structure diverge from all the other players in space travel. Commercial space companies focus on making a profit and on manned

missions that stay inside the solar system. NASA, which also has no plans for interstellar travel, is too risk-averse for something this uncertain; its bureaucratic procedures are often cumbersome and redundant; and its missions are at the mercy of inconsistent congressional approval and funding. "NASA has to take time; billionaires can just do it," says Leroy Chiao, a former astronaut and commander of the International Space Station. "You put this team together, and off you go."

The Game Plan

The man driving the Starshot project has always been inspired by the far reaches. Yuri Milner was born in Moscow in 1961, the same year Yuri Gagarin became the first human to go into space. "My parents sent me a message when they called me Yuri," he says — that is, he was supposed to go somewhere that no one had ever been. So he went into physics — "It was my first love," he says. Milner spent 10 years getting educated, then worked on quantum chromodynamics. "Unfortunately, I did not do very well," he says. Next he went into business, became an early investor in Facebook and Twitter, and amassed a fortune reported to be nearly $3 billion. "So maybe four years ago," Milner says, "I started to think again about my first love."

In 2013 he set up the Breakthrough Prizes, one each for the life sciences, mathematics, and physics. And in 2015 he started what he calls his hobby, the Breakthrough Initiatives, a kind of outreach to the universe: a $1 million prize for the best message to an extraterrestrial civilization, $100 million for a wider, more sensitive search for extraterrestrial intelligence, and now $100 million to Starshot.

In early 2015 Milner recruited a central management team for Starshot from people he had met at various Breakthrough gatherings. Starshot's advisory committee chair and executive director, respectively, are Avi Loeb, chair of Harvard University's astronomy department, and Pete Worden, who directed the NASA Ames Research Center and was involved in a DARPA/NASA plan for a starship to be launched in 100 years. Worden recruited Pete Klupar, an engineer who had been in and out of the aerospace industry and had worked for him at Ames, as Starshot's director of engineering. They in turn pulled together the impressive committee, which includes specialists in the relevant technologies who are apparently willing to participate for some or no money, as well as big names such as Facebook's Mark Zuckerberg and cosmologist Stephen Hawking. Starshot's management policy seems to be a balance between NASA's hierarchical decision-tree rigor and the Silicon Valley culture of putting a bunch of smart people in a room, giving them a long-term goal, and standing back. One committee member, James Benford, president of Microwave Sciences, says the charge is to "give us next week and five years from now, and we'll figure out how to connect the two."

The assembled team members began by agreeing that they could rule out sending humans to Alpha Centauri as too far-fetched and planned to focus on an unmanned mission, which they estimated they could launch in roughly 20 years. They then agreed that the big problem was spacecraft propulsion. So in mid-2015 Loeb's postdocs and graduate students began sorting the options into the impossible, the improbable, and the feasible. In December of that year they received a paper by Philip Lubin, a physicist at the University of California, Santa Barbara, called "A Roadmap to Interstellar Flight." Lubin's option for propulsion was a laser-phased array—that is, a large number of small lasers ganged together so that their light would combine coherently into a single beam. The laser beam would push a sail-carried chip that would need to move at a good fraction of light speed to reach another star within a couple of decades. (A similar idea had been published 30 years earlier by a physicist and science fiction writer named Robert Forward; he called it a Starwisp.) Although the technology was still more science fiction than fact, "I basically handed Starshot the road map," Lubin says, and he joined the project.

In January 2016, Milner, Worden, Klupar, Loeb, and Lubin met at Milner's house in Silicon Valley and put together a strategy. "Yuri comes in, holding a paper with sticky notes on it," Lubin says, "and starts asking the right science and economic questions." The beauty of the project's unusual approach was that rather than going through a drawn-out process of soliciting and reviewing proposals, as NASA would, or being concerned about the potential for profit like a commercial company, the Starshot team was free to hash out a basic plan based purely on what sounded best to it.

Starshot's only really expensive element was the laser; the sails and chips would be low-cost and expendable. The latter would be bundled into a launcher, sent above the atmosphere, and released like flying fish, one after another—hundreds or thousands of them—so many that like the reptilian reproduction strategy, losing a few would not matter. Each one would get hit by the laser and accelerated to 20 percent the speed of light in a few minutes. Next the laser would cut off, and the chip and sail would just fly. When they got to the star, the chips would call back home. "Ten years ago we couldn't have had a serious conversation about this," Milner says. But now, what with lasers and chips improving exponentially and scientists designing and building new materials, "it's not centuries away, it's dozens of years away."

Starshot management sent the idea out for review, asking scientists to look for deal-breakers. None found any. "I can tell you why it's hard and why it's expensive," Lubin says, "but I can't tell you why it can't be done." By April 2016 the team had agreed on the system, and on April 12 Milner arranged a press conference atop the new Freedom Tower in New York City, featuring videos, animations, and several members of the advisory committee. He announced an "interstellar sailboat" driven by a wind of light. The researchers spent the following summer outlining what had to happen next.

StarChips and Light Sails

The team soon found that, though technically feasible, the plan would be an uphill climb. Even the easiest of the technologies, the StarChip, poses a lot of problems. It needs to be tiny—roughly gram-scale—yet able to collect and send back data, carry its own power supply, and survive the long journey. Several years ago engineer Mason Peck's group at Cornell University built what they call Sprites, smartphonelike chips that carry a light sensor, solar panels, and a radio and weigh four grams each. The Starshot chips would be modeled on the Sprites but would weigh even less, around a gram, and carry four cameras apiece. Instead of heavy lenses for focusing, one option is to place a tiny diffraction grating called a planar Fourier capture array over the light sensor to break the incoming light into wavelengths that can be reconstructed later by a computer to any focal depth. Other equipment suggested for the chip include a spectrograph to identify the chemistry of a planet's atmosphere and a magnetometer to measure a star's magnetic field.

The chips would also need to send their pictures back over interstellar distances. Satellites currently use single-watt diode lasers to send information, but over shorter distances: So far, Peck says, the longest distance has been from the moon, more than 100 million times closer than Alpha Centauri. To target Earth from the star, the laser's aim would need to be extraordinarily precise. Yet during the four-year trip the signal will spread out and dilute until, when it reaches us, it will come in as just a few hundred photons. A possible solution would be to send the pictures back by relay, from one StarChip to a series of them flying at regular distances behind. Getting the information back to Earth, says Starshot Advisory Committee member Zac Manchester of Harvard, "is still a really hard problem."

The chips also need batteries to run the cameras and onboard computers to transmit data back during the 20-year voyage. Given the distance to Alpha or Proxima Centauri and the few watts achievable on a small chip, the signal would arrive on Earth weak but "with just enough photons for Starshot's receiver to pick it up," Peck says. To date, no power source simultaneously works in the dark and the cold, weighs less than a gram, and has enough power. "Power is the hardest problem on the chip," Peck says. One possible solution, he offers, is to adapt the tiny nuclear batteries used in medical implants. Another is to tap the energy the sail gains as it travels through the gas- and dust-filled interstellar medium and heats up via friction.

The same interstellar medium could also pose hazards for the Starshot chips. The medium is like highly rarefied cigarette smoke, says Bruce Draine, an astronomer at Princeton University who is also a committee member. No one knows exactly how dense the medium is or what size the dust grains are, so its potential for devastation is hard to estimate. Collisions near the speed of light between the StarChips and grains of any size could create damage that would range from minor craters to complete destruction. If the StarChips are a square centimeter, Draine says, "you'll collide with many, many of these things" along the way. One protectant against smaller particles might be a coating of a couple of millimeters of beryllium copper, although dust grains could still cause catastrophic damage. "The chip will either survive or it won't," Peck says, but with luck, out of the hundreds or thousands sent off in the chip swarm, some will make it.

The next hardest technology is the sail. The StarChips would be propelled by the recoil from light reflected off their sails, the way the recoil from a tennis ball pushes a racket. The more light gets reflected, the harder the push and the faster the sail; to get to 20 percent of light speed, the Starshot light sail has to be 99.999 percent reflective. "Any light that isn't reflected ends up heating the sail," says Geoffrey Landis, a scientist at the NASA Glenn Research Center and a member of the advisory committee—and given the extraordinary temperatures of the light beamer, "even a small fraction of the laser power heating the sail would be disastrous." Compared with today's solar sails, which have used light from the sun to propel a few experimental spacecraft around the solar system, it also has to be much lighter, of a thickness measured in atoms or about "the thickness of a soap bubble," Landis says. In 2000, in the closest approximation yet, Benford used a microwave beam to accelerate a sail made of a carbon sheet. His test achieved about 13 g's (13 times the acceleration felt on Earth caused by gravity), whereas Starshot's sail would need to withstand an acceleration up to 60,000 g's. The sail, like the StarChip, would also have to stand up to dust in the interstellar medium punching holes in it. So far no material exists that is light, strong, reflective, and heat-resistant and that does not cost many millions of dollars. "One of the several miracles we'll have to invent is the sail material," Klupar says.

Other sail-related decisions remain. The sail could attach to the chip with cables, or the chip could be mounted on the sail. The sail might spin, allowing it to stay centered on the light beamer. After the initial acceleration, the sail could fold up like an umbrella, making it less vulnerable during the journey. And once it got to Alpha Centauri, it could unfold and adjust its curvature to act like a telescope mirror or an antenna to send the chip's messages back to Earth. "It sounds like a lot of work," Landis says, "but we've solved hard problems before."

Yet all these challenges are still easier than those of the light beamer that will push the sail. The only way Starshot could reach a good fraction of light speed is with an unusually powerful 100-gigawatt laser. The Department of Defense has produced lasers more powerful, says Robert Peterkin, chief scientist at the Directed

Energy Directorate at the U.S. Air Force Research Laboratory, but they shine for only billionths or trillionths of a second. The Starshot light beamer would have to stay on each sail for minutes. To reach this kind of power for that long, small fiber lasers can be grouped into an array and phased together so that all their light combines into one coherent beam. The Defense Department has also built phased array lasers, but theirs include 21 lasers in an array no more than 30 centimeters across, Peterkin says, which achieves a few tens of kilowatts. The Starshot light beamer would have to include 100 million such kilowatt-scale lasers, and the array would spread a kilometer on each side. "How beyond the state of the art is that?" Peterkin says.

"And it all gets worse and worse," he adds. The 100 million little lasers would be deflected by the normal turbulence of the atmosphere, each one in its own way. In the end the light beamer would need to bring them all to a single focus 60,000 kilometers up on a four-square-meter sail. "At the moment," says Robert Fugate, a retired scientist at the Directed Energy Directorate who is on the committee, drily, "phasing 100 million lasers through atmospheric turbulence on a meter-class target 60 megameters away has my attention." The light could miss the sail completely or more likely hit it unevenly so parts of the sail would be pushed harder, causing it to tumble, spin, or slip off the beam.

Again, the Starshot team has a potential solution but one that comes with its own set of problems. A technology called adaptive optics, already used by large telescopes, cancels out the distortion created by the atmosphere's turbulence with a flexible mirror that creates an equal and opposite distortion. But this technology would need major adaptations to work for Starshot. In the case of the beamer, instead of an adjustable mirror, scientists would have to minutely adjust each laser fiber to make the atmospheric correction. Current adaptive optics on telescopes can resolve at best a point 30 milliarcseconds across (a measure of an object's angular size on the sky). Starshot would need to focus the beamer within 0.3 milliarcsecond across—something that has never been done before.

And even if all these disparate and challenging technologies could be built, they must still work together as a single system, which for the Starshot managers is like creating a puzzle with pieces whose shapes evolve or do not yet exist. Worden calls the process "the art of a long-term hard-research program." The system has "no single design yet," says Kevin Parkin of Parkin Research, a systems engineer who is on the committee. The plan for the first five years, Klupar says, is to "harvest the technologies"—that is, with the guidance of the relevant experts on the committee, the team members will carry out small-scale experiments and make mathematical models. They began in the winter of 2015–2016 by scoping out existing technologies and requesting proposals for not-yet-developed technologies; in spring 2017 they intend to award small contracts of several hundred thousand to $1.5 million each. Prototypes would come next, and, assuming their success, construction of the laser and sail could begin in the early 2030s, with launch in the mid-2040s. By that time Starshot will likely have cost billions of dollars and, with any luck, have collected collaborators in governments, labs, and space agencies in the U.S., Europe, and Asia. "I will make the case, and I hope more people will join," Milner says. "It has to be global," he adds, citing the reasonable national security concerns of an enormous laser installation. "If you start something like this in secrecy, there will be many more question marks. It's important to announce intentions openly."

Starward, Ho!

Given all these hurdles, what are the odds of success? Technologically savvy people not connected to Starshot estimate they are small; several people told me flatly, "They're not going to Alpha Centauri." David Charbonneau of the Harvard-Smithsonian Center for Astrophysics says the project will ultimately be so expensive that "it may amount to convincing the U.S. population to put 5 percent of the national budget—the same fraction as the Apollo program—into it."

Those connected with Starshot think the odds are better but are pragmatic. "We can certainly use lasers to send craft to Alpha Centauri," says Greg Matloff of the New York City College of Technology, a member of the committee. "Whether we can get them there over the next 20 years, I don't know." Harvard's Manchester says, "Within 50 years the odds are pretty good; in a century, 100 percent." Worden thinks their approach is purposefully measured, "and maybe in five years we'll find we can't do it." Milner sees his job on Starshot, besides funding it, as keeping it practical and grounded. "If it takes more than a generation," he says, "we shouldn't work on that project."

Until late last August I thought Dyson was right; the Starshot technology was intriguing, but Alpha Centauri was silly. The star is a binary system (Alpha Centauri A and B), and both stars are sunlike, neither one unusual. Astronomers' understanding of such stars, Charbonneau says, "is pretty good," and although comparing their flares and magnetic fields with our sun's might be useful, "what we'd learn about stellar physics by going there isn't worth the investment."

Now that astronomers know Alpha Centauri's neighbor has a planet, the science case is more promising. The star, Proxima Centauri, is a tad nearer to Earth and is a red dwarf, the most common kind of star. The planet, Proxima Centauri b, is at a distance from its star that could make it habitable. When the discovery was announced, the Starshot team celebrated over dinner. Would members consider changing the project's target? "Sure," Milner says. "We have plenty of time to decide." The laser array should have enough flexibility in pointing that it could "accommodate the difference, about two degrees," Fugate says.

Ultimately the Breakthrough Initiatives' general goal is to find all the planets in the solar neighborhood, Klupar says, and Proxima Centauri b might be just the first. "I feel like an entomologist who picks up one rock, finds a bug, then thinks every rock after that will have a bug under it too," he says. "It's not true, but it's encouraging somehow."

Of course, even the presence of Proxima Centauri b still does not make Starshot slam-dunk science. The chip could take images, maybe look at the planet's magnetic field, perhaps sample the atmosphere—but it would do this all on the fly in minutes. Given the time to launch and the eventual price, says Princeton astrophysicist David Spergel, "we could build a 12- to 15-meter optical telescope in space, look at the planet for months, and get much more information than a rapid flyby could."

But billionaires are free to invest in whatever they wish, and kindred souls are free to join them in that wish. Furthermore, even those who question Starshot's scientific value often support it anyway, because in developing the technology, its engineers will almost certainly come up with something interesting. "They won't solve all the problems, but they'll solve one or two," Spergel says. And an inventive solution to just one difficult problem "would be a great success." Plus, even if Starshot does not succeed, missions capitalizing on the technologies it develops could reach some important destinations both within and beyond our solar system.

Milner's own fondness for the project stems from his hope that it can unite the world's humans in a sense of being one planet and one species. "In the past six years I've spent 50 percent of my time on the road, a lot of time in Asia and Europe," he says. "I realized that global consensus is difficult but not impossible." That theme fits with the other Breakthrough Initiatives, which chiefly want to find aliens to talk to, and with Milner's considerable investments in the Internet and social media, which have changed the nature of conversation and community. But in the end, even he acknowledges that wanting to go to a star is inexplicable. "If you keep asking me why, eventually I'll say I don't know. I just think it's important."

Almost everyone I asked said the same: they cannot explain it to someone who does not already understand—they just want to go. James Gunn, emeritus professor in Princeton's Department of Astrophysical Sciences, who thinks Starshot's chances of success are slim and who dismissed the scientific motivations, still says, "I'm rational about most things, but I'm not particularly rational about the far reach of humanity. I dreamed of going to the stars since I was a kid." Many of the advisory committee said the same thing. "It is just so cool," Landis says, echoing the exact words of other members.

The contradictions inherent in such dreams are perhaps best expressed by Freeman Dyson. Starshot's laser-driven sail with its chip makes sense, he says, and those behind the project are smart and "quite sensible." But he thinks they should stop trying to go to Alpha or Proxima Centauri and focus on exploring the solar system, where StarChips could be driven by more feasible, less powerful lasers and travel at lower speeds. "Exploring is something humans are designed for," he says. "It's something we're very good at." He thinks "automatic machines" should explore the universe —that there is no scientific justification for sending people. And then, being Dyson and unpredictable, he adds, "On the other hand, I still would love to go."

AZEEN GHORAYSHI

He Fell in Love with His Grad Student — Then Fired Her for It

FROM *BuzzFeed News*

CHRISTIAN OTT, A young astrophysics professor at the California Institute of Technology, fell in love with one of his graduate students and then fired her because of his feelings, according to a recent university investigation. Twenty-one months of intimate online chats, obtained by *BuzzFeed News*, confirm that he confessed his actions to another female graduate student.

The university investigation, which concluded in September, found that Ott violated the school's harassment policies with both women. Ott, a 38-year-old rising star who had been granted tenure the year before, was placed on nine months of unpaid leave. During that time he is barred from campus, his communication with most of his postdoctoral fellows will be monitored, and, with the exception of a single graduate student, he is not allowed to have contact with any other students. Before returning he must undergo what a school official calls "rehabilitative" training.

The sanctions were imposed quietly, but after an inquiry from *BuzzFeed News* about Ott's case, the university's president and provost emailed a statement to the entire university on January 4.

"There was unambiguous gender-based harassment of both graduate students by the faculty member," the statement said. It also noted that the faculty member—who was not named—had appealed the sanctions against him, but the university denied his request.

Ott declined to address most questions about his case, telling *BuzzFeed News* he was "constrained from commenting on the situation at this time." But he challenged the idea that he was responsible for anyone's firing.

"At Caltech graduate students are not 'fired' by the decision of a single faculty member," he wrote in an email this week. "When problems with students arise, multiple faculty get involved and a solution is found that ensures the graduate student is not harmed."

Now the two women who filed the harassment complaint—the graduate students Io Kleiser, whom he fell in love with, and Sarah Gossan, whom he confessed his feelings to—have shared their stories with *BuzzFeed News*. They said they were disappointed that instead of terminating Ott's employment, Caltech chose to take a rehabilitative approach and will allow Ott to continue to work with students.

"Because Christian still has a place at Caltech, I feel that I don't," Kleiser, who left the university in January, told *BuzzFeed News*. (Kleiser will finish her research at the University of California, Berkeley but will still receive her doctorate degree from Caltech.) "If they retain Christian and keep a place for him, then they may be inadvertently telling many students that those students do not have a place at Caltech."

Interviews with a dozen of Ott's current and former colleagues, as well as more than 1,000 pages of correspondence between Ott and the two complainants that were submitted to the investigators, suggest that Ott struggled not only with romantic feelings for his student but with forging appropriate professional relationships with some of the people he advised.

Speaking on behalf of the university, Fiona Harrison, chair of the Division of Physics, Mathematics, and Astronomy, told *BuzzFeed News* that the sanctions were appropriately severe. Ott committed "gender-based harassment and discrimination, and we have zero tolerance for that here at Caltech," she said. "I think our actions actually demonstrate that."

Ott's case coincides with high-profile incidents of sexual harassment in university science departments. In October, *BuzzFeed News* revealed that Berkeley had found that the famous astronomer Geoff Marcy had sexually harassed students. And on Tuesday, during a speech on sexism and science on the House floor, congresswoman Jackie Speier of California revealed that a 2004 report from the University of Arizona found that the astronomer Tim Slater had violated sexual harassment policies.

"Science students go to college to study astronomy, chemistry, or physics, not their professors' sex lives," Speier told *BuzzFeed News* by email. "Sexual harassment in science is pervasive," she said, and "the culture needs to change if we want women in this country to reach their full potential as scientists."

Unlike Berkeley, which did not punish Marcy, Caltech has been applauded by some observers for imposing sanctions against Ott. But others question why, despite warning signs, no action was taken until Gossan came forward last June.

"He's obviously a talented researcher, but that's not all that his job entails," said Joan Schmelz, who until recently led the American Astronomical Society's Committee on the Status of Women in Astronomy. "In his current state, should he be advising students and postdocs? I think no."

Io Kleiser came to Caltech to work with Ott in 2012, when she was 22, to study supernovae, the rare astronomical explosions that happen in the final stages of a massive star's life.

Halfway through the year Kleiser was taking a full load of classes as well as doing research with Ott, who uses supercomputers to model the mysterious explosions. She struggled with the workload. "I was just trying to keep my head above water," she said.

Ott began messaging her late at night online, where they talked about their shared insecurities about work. Sometimes their chats were casual; he'd recommend that she read Charles Bukowski or listen to Leonard Cohen. But other times he'd ask her why she wasn't devoting more time to research, questioning her motivations and time management. "It saddens me that research is coming last," he wrote one night in May 2013.

"Not only was he being demanding in terms of my time," Kleiser said, "but he was questioning my commitment to the work, and telling me about how it was making him feel, really from an emotional angle."

In fall 2013, Kleiser went to the incoming executive officer for astronomy, Sterl Phinney, to tell him that she was struggling to work well with her adviser. Within a few weeks, Kleiser said, Ott asked her to meet.

Over coffees at a Peet's just off campus, Kleiser recalled, Ott broke the news that he no longer wanted to work with her, meaning she would have to find another adviser to finish her graduate studies. The change totally upended her research plans, but she said she didn't really understand why he was firing her. He mentioned an email that she had not responded to a few weeks earlier, she recalled, and said he "couldn't emotionally deal with" her anymore. She had no idea that he had any romantic feelings for her. She just thought she had failed at her job.

Five days later, around 1 a.m., Ott messaged her online. "Of all my students I cared most about you and I failed in the worst way," Ott typed. "My problem is that I don't want to be in a power position, but I factually am."

Around the same time Ott began chatting online with another of his female graduate students, 23-year-old Sarah Gossan, to confide in her about the situation with Kleiser. One evening Ott asked Gossan to switch from chat to Skype. "I can't even write this stuff down," he typed.

On Skype a few minutes later, according to Gossan, Ott confessed to being in love with Kleiser. "The reason he had fired her was because he was concerned she was using her sexual influence over him to not do any work," Gossan told *BuzzFeed News*.

Over the next year and a half, Ott continued to message Gossan online, sometimes late at night or while he was inebriated. He talked to her about not being able to let go of his feelings for Kleiser, whom he was still repeatedly reaching out to by chat and email. He also discussed his previous relationships and past emotional involvement with students.

Gossan was often sympathetic to Ott, and opened up to him about her own struggles with anxiety, bulimia, and her boyfriend.

"I am just so happy that I have a female grad student who is actually sane and I can talk to," Ott wrote to her in January 2014.

"Do you think I am a shady person because I let myself be emotionally involved with my student?" he asked her later that month. "I think I may actually be prone to this sort of thing."

Almost immediately after these conversations began, Gossan said, she felt emotionally distraught by them, often working on her supernova research at home rather than at the office, and switching her chat settings to "invisible."

"It's not good if a person in power is out of their fucking mind,"

Ott wrote to her in December 2014, referring to an issue with another student.

"Well we are all out of our minds," Gossan replied.

"Yeah, but your insanity does not affect other people's lifes," he said.

By Gossan's third year Ott's demands on her intensified, she said. "When I said I couldn't work 80 hours a week, he said I would never make it in academia," Gossan recalled. "I came to Caltech to do science. He slowly but surely made me feel worthless."

In April last year, she said, she realized that her deteriorating relationship with Ott was harming her work and emotional well-being. After a dispute in which Ott said she had "not published anything substantial" enough to speak at a conference commemorating Einstein, Gossan reached her breaking point. Two days later she switched advisers, and about a month after that she filed a complaint with Caltech's Title IX office, which handles issues of gender equity.

By that time Kleiser had gotten a new adviser too. But she was still upset about her unexplained firing—she felt like an outcast, she said, uncertain of her place at Caltech or as a scientist. "I went into a several-month-long state of depression where I couldn't even sit down at my computer and work," she said. "It made me feel sick."

Kleiser said she didn't find out about Ott's feelings for her until June 4, when Caltech's Title IX coordinator called her into her office and presented her with a stack of 86 poems Ott had posted about her on his Tumblr page. (The poems, which *BuzzFeed News* has reviewed, are no longer online.) The coordinator told Kleiser she could join Gossan's official complaint.

The two graduate students knew each other, and that night they met for a drink. Kleiser emailed the Title IX office from the bar. "Add my name. Talked to sarah. I am so mad," she wrote. "I will do whatever it takes."

In a letter sent to Kleiser in September, the university acknowledged that her firing "was prompted by [Ott's] romantic or sexual feelings for you" and that his behavior "significantly and adversely affected your educational opportunities at Caltech." A letter sent to Gossan concluded that Ott's interactions with her "placed an inappropriate and undue burden on you that adversely affected your emotional and physical well-being."

In addition to Kleiser and Gossan, seven other students have left Ott's research group since 2012. All of them spoke with *BuzzFeed News*. Four said they were fired abruptly. Many said that Ott's erratic behavior created a hostile and demanding work environment where bullying was the norm.

Casey Handmer was a grad student in Ott's group until June 2013, when he was fired partly because Ott didn't want him to keep his bicycle locked up inside. "Either you accept my rules or you go look for another advisor," Ott wrote him by email. "Your call!"

"As his student, did I have an obligation to manage his moods and pussyfoot my way around the extent to which a grown man is unable to control himself?" Handmer told *BuzzFeed News*. "I hadn't come to Caltech to join some weird cult where you have to do whatever the leader says."

Five other students, including Gossan, quit his group on their own for a variety of reasons, some of which were unrelated to his behavior. Since Ott joined Caltech's faculty in 2009, just two of his graduate students have completed their degrees.

"At this point it's not isolated incidents, it's a statistic," a former student who testified in the Ott case told *BuzzFeed News*. Though she never felt personally harassed, she said she left the research group in part because she found the atmosphere toxic, with Ott often berating and belittling his students.

"It's normal to have an ebb and flow, for students to quietly migrate somewhere else, but having nine students [leave] and two graduate is very strange," Chiara Mingarelli, an astrophysics postdoc at Caltech who works in a different research group and testified in the investigation, told *BuzzFeed News*. "If this was a normal company, there would be no question about his dismissal."

The case underscores a common problem in academia: professors, promoted for their research, may be unequipped to advise students.

"We don't talk enough about how to talk about, and live within, honorable professional boundaries in the role of professor," C. K. Gunsalus, director of the National Center for Professional and Research Ethics, told *BuzzFeed News*. (Gunsalus was not involved with

Ott's case.) "There isn't always as much formal preparation for the teaching and advising role as is needed for people."

Harrison, the division chair, told *BuzzFeed News* that Caltech considers the success rate for graduating students when a professor is up for tenure, though she would not speak to how it was weighed in Ott's case.

Caltech is planning to offer more mentorship training for junior faculty, Harrison added, and is considering how to get confidential feedback from students about how professors advise them. "We are drawing every lesson we can from what happened."

In the first week of January, Kleiser moved to Berkeley to continue her graduate work on supernovae. Caltech did not tell her or Gossan that it would be sending out a university-wide statement about their former adviser and the complaint they played such a big role in.

"I did not know it would go out, but I am dealing with it fine," Kleiser wrote in an email. "I just moved out of my apartment an hour ago and am driving up to Berkeley tomorrow, so I am thinking about other stuff. :)"

Gossan, however, is not feeling as settled. She will stay at Caltech to finish her degree, which she expects to get next year. Ott will be allowed back on campus July 1.

CHRIS JONES

The Woman Who Might
Find Us Another Earth

FROM *The New York Times Magazine*

LIKE MANY ASTROPHYSICISTS, Sara Seager sometimes has a problem with her perception of scale. Knowing that there are hundreds of billions of galaxies and that each might contain hundreds of billions of stars can make the lives of astrophysicists and even those closest to them seem insignificant. Their work can also, paradoxically, bolster their sense of themselves. Believing that you alone might answer the question "Are we alone?" requires considerable ego. Astrophysicists are forever toggling between feelings of bigness and smallness, of hubris and humility, depending on whether they're looking out or within.

One perfect blue-sky fall day, Seager boarded a train in Concord, Massachusetts, on her way to her office at MIT and realized she didn't have her phone. She couldn't seem to decide whether this was or wasn't a big deal. Not having her phone would make the day tricky in some ways, because her sons, 13-year-old Max and 11-year-old Alex, had a soccer game after school, and she would need to coordinate a ride to watch them. She also wanted to be able to find and sit with her best friend, Melissa, who sometimes takes the same train to work. "She's my best friend, but I know she has other best friends," Seager said, wanting to make the nature of their relationship clear. She is an admirer of clarity. She also likes absolutes, wide-open spaces, and time to think, but not too much time to think. She took out her laptop to see if she could email

Melissa. The train's Wi-Fi was down. She would have to occupy herself on the commute alone.

Seager's office is on the 17th floor of MIT's Green Building, the tallest building in Cambridge, its roof dotted with meteorological and radar equipment. She is a tenured professor of physics and of planetary science, certified a "genius" by the MacArthur Foundation in 2013. Her area of expertise is the relatively new field of exoplanets: planets that orbit stars other than our sun. More particularly, she wants to find an Earthlike exoplanet—a rocky planet of reasonable mass that orbits its star within a temperate "Goldilocks zone" that is not too hot or too cold, which would allow water to remain liquid—and determine that there is life on it. That is as simple as her math gets.

Her office is spare. There is a set of bookshelves—*Optics* and *Asteroids III* and *How to Build a Habitable Planet*—topped with a row of certificates and honors leaning against a chalkboard covered with equations. In addition to the MacArthur award, which doesn't come with a certificate but with $625,000, she is proudest of her election to the National Academy of Sciences. Although the line between lunacy and scientific fact is constantly shifting, the search for aliens still occupies the shadows of cranks, and Seager hears from them almost daily, or at least her assistant does. By the standards of her universe, Seager is famous. She is careful about the company she keeps and the words she chooses. She isn't searching for aliens. She's searching for exoplanets that show signs of life. She's searching for a familiar blue dot in the sky.

That means Seager, who is 45, has given herself a very difficult problem to solve, the problem that has always plagued astronomy, which, at its essence, is the study of light: Light wages war with itself. Light pollutes. Light blinds.

Seager has a commanding view of downtown Boston from her office window. She can sweep her eyes, hazel and intense, all the way from the gold Capitol dome to Fenway Park. When Seager works at night and the Red Sox are in town, she sometimes has to close her curtains because the ballpark's white lights are so glaring. And on this morning, after the sun completed its rise, her enviable vista became unbearable. It was searing, and she had to draw her curtains. That's how light can be the object of her passion and also her enemy. Little lights—exoplanets—are washed out by bigger lights—their stars—the way stars are washed out by our biggest light, the sun. Seager's challenge is that she has dedicated her life to the search for the smallest lights.

The vastness of space almost defies conventional measures of distance. Driving the speed limit to Alpha Centauri, the nearest star grouping to the sun, would take 50 million years or so; our fastest current spacecraft would make the trip in a relatively brisk 73,000 years. The next nearest star is six light-years away. To rocket across our galaxy would take about 23,000 times as long as a trip to Alpha Centauri, or 1.7 billion years, and the Milky Way is just one of hundreds of billions of galaxies. The Hubble Space Telescope once searched a tiny fragment of the night sky, the size of a penny held at arm's length, that was long thought by astronomers to be dark. It contained 3,000 previously unseen points of light. Not 3,000 new stars—3,000 new galaxies. And in all those galaxies, orbiting around some large percentage of each of their virtually countless stars: planets. Planets like Neptune, planets like Mercury, planets like Earth.

As late as the 1990s, exoplanets remained a largely theoretical construct. Logic dictated that they must be out there, but proof of their existence remained as out of reach as they were. Some scientists dismissed efforts to find exoplanets as "stamp collecting," a derogatory term within the community for hunting new, unreachable lights just to name them. (Even among astronomers there can be too much stargazing.) It wasn't until 1995 that the colossal 51 Pegasi b, the first widely recognized exoplanet orbiting a sunlike star, was found by a pair of Swiss astronomers using a light-analyzing spectrograph. The Swiss didn't see 51 Pegasi b; no one has. By using a complex mathematical method called radial velocity, they witnessed the planet's gravitational effect on its star and deduced that it must be there.

There has been an explosion of knowledge in the relatively short time since, in part because of Seager's pioneering theoretical work in using light to study the composition of alien atmospheres. When starlight passes through a planet's atmosphere, certain potentially life-betraying gases, like oxygen, will block particular wavelengths of light. It's a way of seeing something by looking for what's not there.

Light or its absence is also the root of something called the transit technique, a newer, more efficient way than radial velocity of finding exoplanets by looking at their stars. It treats light almost like music, something that can be sensed more accurately than it can be seen. The Kepler space telescope, launched in 2009 and now trailing 75 million miles behind Earth, detects exoplanets when they orbit between their stars and the telescope's mirrors, making tiny but measurable partial eclipses. A planet the size of Jupiter passing in front of its sun might result in a 1 percent dip in the amount of starlight Kepler receives, a drop that in time reveals itself to be as regular as rhythm, as an orbit. The transit technique has led to a bonanza of finds. In May, NASA announced the validation of 1,284 exoplanets, by far the largest single collection of new worlds yet. There are now 3,414 confirmed exoplanets and an additional 4,696 suspected ones, the count forever increasing.

Before Kepler, the nature of the transit technique meant that most of those exoplanets were "Hot Jupiters," giant balls of hydrogen and helium with short orbits, making them scalding, lifeless behemoths. But in April 2014, Kepler found its first Earth-size exoplanet in its star's habitable zone: Kepler-186f. It's about 10 percent larger than Earth and orbits on the outer reaches of where the temperature could allow life. No one knows the mass, composition, or density of Kepler-186f, but its discovery remains a revelation. Kepler was searching, somewhat blindly, an impossibly small sliver of space, and it found a potentially habitable world more quickly than anyone might have guessed.

In August astronomers at the European Southern Observatory announced that they had detected a somewhat similar planet orbiting Proxima Centauri, the single star closest to us after the sun. They named it Proxima Centauri b. Studying the data, Seager supported the discovery and agreed that it might boast a life-sustaining—or at least non-life-threatening—surface temperature. There are now nearly 300 confirmed exoplanets or candidates orbiting within the habitable zones of their stars. Extrapolating the math, NASA scientists now believe that there are tens of billions of potentially life-sustaining planets in the Milky Way alone. The odds practically guarantee that a habitable planet is somewhere out there and that someone or something else is too.

In some ways the search for life is now where the search for exoplanets was 20 years ago: common sense suggests a presence that we can't confirm. Seager understands that we won't know they're out there until we more truly lay eyes on their home and see something that reminds us of ours. Maybe it's the color blue; maybe it's clouds; maybe, however many generations from now, it's the orange electrical grids of alien cities, the black rectangles of their lightless Central Parks. But how could we ever begin to look that far? "Everything brave has to start somewhere," Seager says.

The beginning of her next potential breakthrough hangs on the wall opposite the window in her office. It is a two-thirds scale model of a single petal of something called the starshade. She has been a leading proponent of the starshade project, and outside her teaching, it is one of her principal professional concerns.

Imagine that far-off aliens with our present technology were trying to find us. At best, they would see Jupiter. We would be lost in the sun's glare. The same is true for our trying to see them. The starshade is a way to block the light from our theoretical twin's sun, an idea floated in 1962 by Lyman Spitzer, who also laid the groundwork for space telescopes like Hubble. The starshade is a huge shield, about 100 feet across. For practical reasons that have to do with the bending of light but also lend it a certain cosmic beauty, the starshade is shaped exactly like a sunflower. By Seager's hopeful reckoning, one day the starshade will be rocketed into space and unfurled, working in tandem with a new space telescope like the WFIRST, scheduled to launch in the mid-2020s. When the telescope is aimed at a particular planetary system, lasers will help align the starshade, floating more than 18,000 miles away, between the telescope and the distant star, closing the curtains on it. With the big light extinguished, the little lights, including a potential Earthlike planet and everything it might represent, will become clear. We will see them.

The trouble is that sometimes the simplest ideas are the most complicated to execute. About once a decade since Spitzer's proposal—he could work out the math but not the mechanics—someone else has taken up the cause, advancing the starshade slightly closer to reality before technological or political inertia sets in. Three years ago Seager joined a new, NASA-sponsored study to try to overcome the final practical hurdles; NASA then chose her from among her fellow committee members to lead the effort.

After those decades of false starts, Seager and her team have already succeeded in making the starshade seem like a real possibility. NASA recognized it as a "technology project," which is astral-bureaucracy-speak for "this might actually happen." Today the starshade is a piece of buildable, functional hardware. Seager packs that single petal into a battered black case and wheels it, along with a miniature model of the starshade, into classrooms and conferences and the halls of Congress, trying to find the momentum and hundreds of millions of dollars that allow impossible things to exist.

"If I want the starshade to succeed, I have to help mastermind it," Seager says. "The world sees me as the one who will find another Earth." She has her intelligence, and her credentials, and her audience. She has her focus. But maybe more than anything else, Seager understands in ways few of us do that sometimes you need darkness to see.

Seager grew up in Toronto, wired in a way all her own. "Ever since I was a child, there was just something about me that wasn't quite like the others," she says. "Kids know how to sort through who's the same and who's different." After her parents divorced, her father, Dr. David Seager, achieved a certain fame by becoming one of the world's leaders in hair transplants. The Seager Hair Transplant Centre still operates and bears his name a decade after his death. David Seager was besotted with his bright daughter and wanted her to become a physician.

Seager did her best to fit in. Sometimes she did; mostly she didn't. Eventually she gave up trying. She still talks breathlessly —"without enough modulation," she has learned by listening to other people talk. She has never had the patience to invest in something like watching TV. "Things just move too slowly," she says. "It feels like a drag." She sleeps a lot, but that's just a concession to her biology; she recognizes that she's a more efficient machine when she's rested. But if Seager's apartness didn't make her insecure, it also made her feel as though the expectations of others didn't apply to her. "I loved the stars," she says. When she was 16, she bought a telescope.

Friendless for most of her childhood, Seager eventually forged her way to her own vision of the good life. She found and married a quiet man named Mike Wevrick, whom she met on a ski trip with her canoe club. He had seen something in her that nobody other than her father fully saw; he saw her as special as well as strange. Later she graduated from Harvard, an early expert in exoplanets. (51 Pegasi b was discovered just when she was searching for a thesis topic. "I was born at the perfect time," she says.) She and Wevrick had Max and Alex; Seager was hired by MIT, and she and Wevrick and the boys moved into a pretty yellow Victorian in Concord, Massachusetts. She took the train to work. Wevrick, a freelance editor, managed just about everything that didn't involve the search for intelligent life in the universe. Seager never shopped for groceries or cooked or pumped gas. All she had to do was find another Earth.

Then, in the fall of 2009, Wevrick got a stomachache that drove him to bed. They figured it was the flu. Wevrick didn't have the flu but a rare cancer of the small intestine. They were told that the initial prospects were good, and he fought the cancer sufferer's systematic fight. But while laws govern astrophysics, cancer is an anarchist. About a year after Wevrick's diagnosis, he and Seager went cross-country skiing, and he couldn't keep up. A few more terrible months passed, and he began writing out a methodical three-page list, practical advice for Seager after his death. It wasn't a love letter; it was an instruction manual for life on Earth. By June 2011 he was 47 and in home hospice. Seager asked him how to get the roof rack that carried his canoes off the car. "It's too complicated to explain," Wevrick said. That July he died.

The first couple of months after Wevrick's death were weird. Seager felt a surprising sense of relief from the uncertainties of sickness, a kind of liberation. She didn't care about conventions like money, which she had never needed to manage, and she took the boys on some epic trips. There are pictures of them smiling together in the deserts of New Mexico, on mountaintops in Hawaii. Then one day she went into Boston for a haircut, got turned around, and accidentally walked into a lawyer's office next to the salon. Seager ended up talking to a woman inside. That woman was also a widow, and she told Seager that there would be a moment, as inevitable as death itself, when her feelings of release would be replaced by the more lasting aimlessness of the lost. Seager walked back outside, and just like that the world came out from under her feet. She fell into an impossible blackness.

Later that winter she took the boys sledding at the big hill in Concord. Two other women and their children were there. Seager stared at them coldly. They were smiling and carefree with their perfect, blissful lives. Seager felt ugly and ruined next to them. Then Alex, who was six at the time, had a meltdown. He sprawled himself across the hill so that the other children couldn't go down it. The two other mothers tried to get him to move. "He has a problem," Seager told them. They continued to try to shift him.

"HE HAS A PROBLEM," Seager said. "MY HUSBAND DIED."

"Mine too," one of the other women said. That was Melissa. A few weeks later, on Valentine's Day, Seager was invited to her first gathering of the widows.

Today Melissa says she could detect the telltale "flintiness" of the recently bereaved the moment she saw Seager on the hill. Now there were six widows united in Concord, each middle-aged, each in a different stage of grief, drawn together by the peculiar pull of the unlucky. Three had been widowed by cancer, two by accidents —bicycling and hiking—and one by suicide. Melissa's husband was four years gone, Seager's seven months.

Widowhood was like a new universe for Seager to explore. She had never understood many social norms. The celebration of birthdays, for instance. "I just don't see the point," she says. "Why would I want to celebrate my birthday? Why on earth would I even care?" She had also drawn a hard line against Christmas and its myths. "I never wanted my kids to believe in Santa." After Wevrick's death she became even more of a satellite, developing a deeper intolerance for life's ordinary concerns.

Making dinner seemed an insurmountable chore, the routine of school lunches a form of torture. The roof needed to be replaced, and she didn't have the faintest idea how to get it fixed. She wasn't sure how to swipe credit cards. If the answers to her questions weren't somewhere on Wevrick's three wrinkled sheets of paper, it could feel as though they were locked in a safe.

There was a pendant light in her front hall, where the boys would fight with their toy light-sabers, and sometimes they would hit the light with their wild swings. Seager decided that either the light or one of the boys was going to end up damaged. She asked the widows how to do electrical work—"I have to parcel out things with logic and evidence," she says—got out the ladder, and took down the light, carefully wrapping black tape around the ends of the bare wires that now poked through the hole in the ceiling. She remembers thinking that her removing that light all by herself represented the height of her new accomplishment. She felt so reduced. She felt so gigantic.

*

For all of her real and perceived strangeness, the most unusual thing about Seager is her blindness to her greatest gift. She is more than aware of her preternatural mathematical abilities, her possession of a rare mind that can see numbers and their functions as clearly as the rest of us see colors and shapes. "I'm good at that stuff," she says with her brand of factual certainty that is sometimes confused with arrogance. She knows she is unusually capable of turning abstract concepts into things that can be packed into a case. What she doesn't always see is her knack for connection between places, if not always people, the unconventional grace she possesses when it comes to closing unfathomable distances.

Seager has lined the hallway outside her office with a series of magical travel posters put out by the Jet Propulsion Laboratory. Each gives a glimpse of the alien worlds that, in part because of her, we now know exist. There's a poster for Kepler-16b, an exoplanet that orbits a pair of stars, like Luke Skywalker's home planet of Tatooine. Kepler-186f is depicted with red grass and red leaves on its trees, because its star is cooler and redder than the sun, which might influence photosynthesis in foliage-altering ways. There's even one for PSO J318.5-22, a rogue planet that doesn't orbit a star but instead wanders across the galaxy, cast in perpetual darkness, swept by rain of molten iron.

After the discovery of Proxima Centauri b, Seager wrote a galactic postcard from it for the website *Quartz*. She closed her eyes and imagined a world 25 trillion miles away. "For the average earthling," she wrote, "visiting this planet might not be much fun." She saw a planet perhaps a third larger than Earth, with an orbit of only 11 days. Given its proximity to its small, red star, she suggested that the ultraviolet radiation on Proxima Centauri b is probably intense but the light Martian-dim. She also deduced that Proxima Centauri b is "tidally locked." Like the moon's relationship to Earth, one side of the planet always faces its star, which is always in the same place in its sky. Parts of Proxima Centauri b are cast in perpetual sunrise or sunset. One side is always in darkness.

At first after Wevrick's death Seager thought about abandoning her work, because she was having such a hard time with her responsibilities at home. Her dean talked her out of quitting, giving her financial support to hire caregivers for the boys and urging her to redouble her efforts. "I had worked so hard," she says. "I had all the years I called the lost years with Mike when I ignored him. We had little tiny kids. I was working all the time, exhausted all the time. But I was like, We'll have money someday. We'll have time someday."

She paused. Her face was blank, emotionless. "Now I'll cry." Seconds later tears spilled out of her eyes, and her voice modulated. "I wanted to make it up to him, and I never did."

Seager has always found comfort and perhaps even solace in her work, in her search for another and maybe better version of our world. In her mourning, each discovery represented one more avenue of escape. In the spring of 2013 she was given responsibility for the starshade. That July she met a tall, fast-walking man named Charles Darrow.

Darrow, who is now 53, was an amateur astronomer and the president of the Toronto branch of the Royal Astronomical Society of Canada, and at the last minute he decided to go to the society's annual meeting in Thunder Bay, Ontario. Darrow was on his way out of a profoundly unhappy marriage; he worked for his family business, an engine-parts wholesaler. He needed a break, and he pointed his car north. "I wanted to be alone," he says. At a reception on the Friday evening, Darrow noticed a hazel-eyed woman staring at him from across the room. "I thought she was looking at someone behind me," he says. Then he went into the lecture hall, and the same woman was that night's keynote speaker. She talked about exoplanets. The next day, lunch was in a university cafeteria. The woman was in the salad line ahead of him, and she turned around. Darrow mustered up his courage and invited Sara Seager to join him. "I knew about five minutes into the conversation that my life was going to change," he says.

Seager was taken with Darrow the night she saw him in Thunder Bay. She had been struck by the contrast between the whiteness of his shirt and his tanned summer skin. But she didn't have the same certainty that possessed him at their lunch the next day. She wasn't sure how to develop a relationship across the 549 miles between her home in Concord and his home outside Toronto. She thought they might never cross paths again.

They might not have, except Darrow resolved during his drive back home that he had to call her. He picked up the phone five times but always hung up before she answered. On the sixth he spoke to her, beginning a long correspondence, emails and conversations over Skype. Darrow and Seager talked every way but face-to-face. They fell in love remotely. "I had to follow my heart," Darrow says. "I decided that I wasn't going to die unhappy."

Melissa, meanwhile, told Seager that if she could close the gap between here and a planet like Kepler-186f—a journey that would take us 500 light-years to complete—then the 549 miles between Concord and Toronto shouldn't seem like such an insurmountable gulf. By her usual measures, he was right next door.

Seager and Darrow married in April 2015. In different ways, each had rescued the other. Seager was the cataclysm that allowed Darrow to make every correction. He divorced, left his family business, and moved into a pretty yellow Victorian in Concord. The two boys started calling him Dad. For Seager, Darrow was a second chance to know love, even deeper than the one she had known, because it seemed so improbable in her sadness. "I feel so lucky to have found him," Seager says. "What are the chances?"

Adapting to his new life hasn't always been easy for Darrow. He is determined, as he puts it, "to make Sara the happiest woman in the multiverse." He cooks dinner; he helps take care of the boys; he maintains the house; he walks with Seager to the train station every morning, and he picks her up every night. He has chosen to take care of the mundane so that she can devote herself to the extraordinary. But he banged his head more than once on Wevrick's canoe, which still hung from the back of the garage.

Not long ago Darrow was looking for the right ways to assert his presence, to make a claim to a house that didn't always feel like his. The wires dangling from the front hall ceiling bothered him. They looked bad and seemed dangerous. A few months after his arrival in Concord, he took his opening. He carved out some of the plaster, installed a plastic box, ran the wires through it, and hooked up a new fixture, flush-mounted, so that the boys wouldn't hit it during their duels.

Darrow climbed down from the ladder and flicked the switch.

The morning after she forgot her phone, Seager woke up and decided, just like that, to skip the commute. With the house to herself, she tried to make coffee. She left out part of the machine, and after some terrible noises the pot was bone-dry. She sat down at her kitchen table with her empty mug and began talking about hundreds of billions of galaxies and their hundreds of billions of stars. Tens of billions of habitable planets, far more of them than there are people on Earth. There has to be other life somewhere out there. We can't be that special.

"It would be arrogant to think so," Seager said. But in her lifetime, after the WFIRST telescope rockets into orbit, and maybe her starshade follows it—she puts the chances of success at 85 percent—she will have time to explore only the nearest hundred stars or so. A hundred stars out of all those lights in the sky, a fraction of a fraction of a fraction.

Will one of them have a small, rocky planet like Earth? Probably. Will one of those small, rocky planets have liquid water on it? Possibly. Will the planet sustain life? Now the odds tilt. Now they are working against her, and she knows it. Now they're maybe one in a million that she'll find what she's looking for.

She did some private math. "I believe," she said.

Seager's discovery will be fate-altering if it comes, but it will also be quiet, a few pixels on a screen. It will obey the laws of physics. It will be a probability equation: *What are the chances?* We won't discover that there is life on other planets the way we've been taught that we'll learn. There won't be some great mother ship descending from the sky over Johannesburg or a bizarre lightning storm that monsters will ride to New Jersey. What Seager will have is a photograph from a space telescope of a distant solar system, with its star eclipsed by her starshade, and with a familiar blue dot some safe and survivable distance away from it. That's all the evidence she will have that we're not alone, and that will be all the evidence she will need. Her proof of life will be a small light where there wasn't one before.

KATHRYN JOYCE

Out Here, No One Can
Hear You Scream

FROM *Huffington Post Highline/The Nation Institute Investigative Fund*

ON AN EARLY Friday morning in late June 2006, Cheyenne
Szydlo, a 33-year-old Arizona wildlife biologist with fiery red hair,
drove to the Grand Canyon's South Rim to meet the river guide
who would be taking her along the 280 miles of the Colorado
River that coursed a mile below. She was excited. Everyone in her
field wanted to work at the Grand Canyon, and after several years
of unsuccessful applications, Szydlo had recently been offered a
seasonal position in one of the National Park Service's science
divisions. She'd quit another job in order to accept, certain her
chance wouldn't come again.

The Grand Canyon is a mecca of biological diversity, home to
species that grow nowhere else on Earth. But after a dam was built
upstream 60 years ago, changes in the Colorado's flow have en-
abled the rise of invasive species and displaced numerous forms
of wildlife. Szydlo's task was to hunt for the southwestern willow
flycatcher, a tiny endangered songbird that historically had nested
on the river but hadn't been seen in three years. Her supervisor
believed the bird was locally extinct, but Szydlo was determined
to find it. The June expedition—a nine-day journey through the
canyon on a 20-foot motorboat operated by a boatman named
Dave Loeffler—would be her last chance that summer. When
Szydlo asked a coworker what Loeffler was like, the reply was cryp-
tic: "You'll see."

Szydlo, who'd studied marine biology in Australia and coral

reefs in French Polynesia, was drawn to the adventurous nature of the work. "From my earliest memories," she told me, "there was never any place that felt safer or happier to me than the outdoors." On the morning of the trip, she arrived at the boat shop early. She assumed they'd leave at once, to make the most of the day. Instead, she said, Loeffler took her to a coworker's house, and for an hour and a half she sat uncomfortably as Loeffler told his friend about the battery-powered blender he'd packed to make "the best margaritas on the river."

They set out from Lees Ferry in Marble Canyon, the otherworldly antechamber to "the Grand." From there the river winds through towering, striated red cliffs and balancing rock formations, under the Navajo Bridge, and, at around mile 60, into the Grand Canyon itself. The views are stupefying, the waters turquoise, and the disconnection almost total—a moonscape beyond cell-phone reception. For many people it's a spiritual experience.

It's also an intimate one. Travelers eat and sleep together, and due to the lack of cover must often bathe and go to the bathroom in full view, using portable metal ammo cans outfitted with toilet seats. Commercial river guides often say that no one can claim their privacy on the river, so fellow passengers should offer it instead.

In Szydlo's recounting of the trip, Loeffler didn't adhere to this code. When she bent to move provisions or tie up the boat, he commented on a logo on the back of her utility skirt. He asked frank questions about her sex life and referred to Szydlo as "hot sexy biologist." That June the temperatures at the bottom of the canyon reached 109 degrees, and when Szydlo scorched her skin on a metal storage box, Loeffler said she had a hot ass. He adjusted her bra strap when it slipped, and one chilly night invited her to sleep in the boat with him if she was cold. When they stopped to take a picture at a particularly scenic spot, he suggested that she pose naked. He told her that another female Park Services staffer would be hiking in to meet them at the halfway point and that he hoped they would have "a three-way." Szydlo told me she laughed uncomfortably and spoke often of her boyfriend and their plans to get married.

By the third day of the trip, it seemed to Szydlo that Loeffler was getting increasingly frustrated. They stopped at a confluence where the Colorado meets a tributary and forms a short tumble of rapids gentle enough for boaters to swim through with a life jacket. Szydlo pulled on her preserver, but Loeffler insisted she didn't need one. When she entered the river without it, the water sucked her under. She somersaulted through the rapids "like I was in a washing machine," she recalled. She thought she was going to drown. Then the rapids spat her out into a calm, shallow pool. She came up gasping and choking to the sound of Loeffler's laughter, and thought to herself, "I'm in deep shit."

We're used to hearing stories of sexual harassment in the army, the navy, or within the police force; 25 years after the Tailhook scandal, when scores of Marine and naval officers allegedly sexually assaulted some 83 women and seven men at a military convention, there's a general cultural understanding of what women face in traditionally male-dominated public institutions. The agencies that protect America's natural heritage enjoy a reputation for a certain benign progressivism—but some of them have their own troubling history of hostility toward women.

In 2012 in Texas, members of the Parks and Wildlife Department complained about a "legacy" of racial and gender intolerance; only 8 percent of the state's 500 game wardens were women. In 2014 in California, female employees of the U.S. Forest Service filed a class-action lawsuit—the fourth in 35 years—over what they described as an egregious, long-standing culture of sexual harassment, disparity in hiring and promotion, and retaliation against those who complained. (That lawsuit is still pending.) And this January the Department of the Interior's Office of Inspector General announced that it had "found evidence of a long-term pattern of sexual harassment and hostile work environment" in the Grand Canyon's River District, a part of the Park Service.

Ever since the U.S. created institutions to protect its wilderness, those agencies have been bound up with a particular image of masculinity. The first park rangers in the U.S. were former cavalrymen, assigned to protect preserves like Yellowstone and Yosemite from poachers and fire. The public quickly became enamored of these rugged, solitary figures. In the early 1900s, as the Park Service was created, a new breed emerged: naturalists who endeavored to teach the public the principles of conservation. As the historian Polly Welts Kaufman has written, the earlier generation of rangers resented the intrusion of "pansy-pickers" and "butterfly chasers." Also controversial was the presence of a small number of women at the agency. Male naturalists worried that their job would be seen as effeminate instead of, as one put it, "the embodiment of Kit Carson, Daniel Boone, the Texas Rangers, and General Pershing." In the 1930s and '40s the ranks were mostly filled by returning veterans attracted by the ranger corps' quasi-military culture. Until 1978 female rangers weren't permitted to wear the same uniform or even the same badge as the men, but instead wore skirts modeled on stewardesses' uniforms.

The other major institution tasked with preserving and managing the American wilderness, the Forest Service, developed on a similar trajectory. Although the Forest Service comes under the direction of the Department of Agriculture (while the Park Service falls under the DOI), its employees perform similar work and its culture is also modeled along military lines. By the 1970s women held only 2 percent of full-time professional roles in the service nationwide. In California—whose lands are the crown jewel of the national forest system—female employees filed a class-action lawsuit known as *Bernardi v. Madigan*. The case was settled in 1981 with a court-enforced "consent decree" that required the Forest Service's California region to employ as many women as the civilian workforce—at least 43 percent in every pay grade. The decision ultimately saw hundreds of "Bernardi women" enter the service, to the disgruntlement of many male employees.

Lesa Donnelly is a former Forest Service administrator who worked for the agency from 1978 to 2002. In 1994 she filed a complaint charging that three of her male colleagues were harassing her. After word spread (incorrectly) that she planned to file a class-action lawsuit, she received dozens of calls. She heard from women who claimed they were being threatened with physical and sexual assault and women who said they'd been punished for making complaints. One said the men on her crew joked about raping her in her sleep and had tied her bloodstained underwear to the antenna of their fire truck. Two women told her that a notice in their office about the Bernardi consent decree had been defaced with a scrawled reference to the "cuntsent decree." She realized her own complaint was "nothing compared to what I found out was happening."

Eventually Donnelly compiled claims from 50 women, and in 1995 she filed a class-action suit against the Forest Service, including declarations from many of the woman who had reached out to her. The agency negotiated a settlement that allowed for continued court oversight of California's Forest Service. But when the monitoring period ended in 2006, the old problems soon resurfaced, as Donnelly would describe in testimony to Congress two years later. One dispatcher reported that she'd been sexually assaulted and stalked by a manager. He was made to resign, but after six months the Forest Service tried to work with him again. In 2008 a male supervisor at the same forest said that he hated a black female employee and wanted to shoot subordinates he hated. When the employee reported the comment, the district ranger told her to ignore him.

This year I met Donnelly, who is 58, in El Dorado Hills, outside Sacramento. Now the vice president of the USDA Coalition of Minority Employees, a civil rights group, she has the demeanor of a friendly bulldog. She told me that nearly every year for the last 15 years she has traveled to Washington, D.C., to lobby the USDA, Congress, and the White House to protect women in the service. She managed to enlist the help of representatives Jackie Speier of California, Peter DeFazio of Oregon, and Raúl Grijalva of Arizona, who in 2014 petitioned the USDA to investigate, without success. Each time Donnelly comes to D.C., she added, she brings details of 20 to 25 new allegations. But while her fight against the Forest Service has persisted for more than two decades, in the Grand Canyon, similar questions about the treatment of women have only started to surface.

"On the river, the boatman is god," Cheyenne Szydlo told me. In the Grand Canyon, river guides enjoy an almost exalted status, revered for their ability to "read water." Boatmen have almost total responsibility for their passengers—they keep the food and determine when and where to sleep, explore, or go to the bathroom. They also control the satellite phone, the only means of contact with the outside world. But within the Park Service, boatmen were more important still. Men like Dave Loeffler guided visiting officials or VIPs on adventures within the canyon, undertook rescue missions, and were featured in travel stories in newspapers and magazines. They "made it seem [to park management] like the river was the surface of Mars," one boatman for a private company recalled. The administration saw them as irreplaceable.

In the early 2000s three men turned the boat shop into a small fiefdom. There were the "two Daves"—Loeffler and his supervisor, Dave Desrosiers—and Bryan Edwards, the boat shop manager. In addition to this small core of permanent staffers, the park periodically hired intermittent boatmen. One, Dan Hall, worked in the canyon during this period and was friendly with the trio. Hall is garrulous and not remotely prudish. "I have offended people I've worked with," he told me. "I do my best to apologize and not let it happen again . . . But with the Daves, it had this very dark side to it." He remembered the three talking about who could sleep with the most women on the river. "They were always on the make," he said. In a written response sent via Facebook, Edwards said that "no competition ever existed."

Rafting on the Colorado has always had a bit of a party vibe, and that attitude held for Park Service trips too. Boats sometimes carried a large quantity of alcohol. Participants sometimes hooked up. But during the early 2000s, Hall told me, it seemed short-lived river affairs were almost expected of female employees. According to one former employee, veteran female staffers warned new hires to make sure they set up tents with a friend rather than sleeping on the boats, as the boatmen usually did. Sometimes, Hall said, boatmen would lobby supervisors to send women from completely unrelated park divisions—an attractive new hire at the entry booth, for instance—on trips. Often, though, the targets were from science divisions that required river access, such as vegetation and wildlife.

The field leader of the vegetation program from 2002 to 2005, Kate Watters, said that she complained to her supervisor about the boatmen's behavior. In October 2005 an expedition was planned to see if the two groups could overcome their difficulties. The trip was led by Bryan Edwards. Participants included Watters, who was married to Dan Hall at the time, and her new intern, a biologist I'll call Anne.

The expedition coincided with Halloween week, and one night most of the participants put on costumes. Many were drinking. Anne—dressed as a butterfly, in wings and a dress—was in the camp's kitchen area when Edwards—dressed as a pirate—came up behind her. He grabbed the camera she'd left on the table. "The next thing I knew, his hand was between my legs," she said. Then Edwards shot a picture up her skirt.

Watters observed aloud that Edwards's behavior was unacceptable. Loeffler, who was attired as "a hillbilly ax murderer" and carrying a real ax, demanded that Watters talk it out with Edwards instead of filing a report. She recalled that he bellowed at her, ax in hand, "Fuck you, Kate Watters. You can't have control over people's jobs." Loeffler told me that he was unable to answer questions since he is still a park employee. Edwards wrote in his response, "I did flash a camera below her skirt as she stood next to me. It was intended for shock value only" as Anne had been drinking, he explained.

Watters said that in a meeting after her return with Edwards's and Desrosiers's boss, Edwards glared at her and cleaned his nails with a six-inch buck knife. (Edwards called this description "entirely false.") In 2006 he received a 30-day suspension over the incident, after which he resigned. Edwards confirmed this to me but wrote in another message, "I suspect nearly everything you have been told is at least either 'misrepresentation' or outright lie." He felt that he had done "a lot of good in my 12 yrs in Grand Canyon," he went on. "Because of my abilities, I did things people dreamed about doing but simply could not on that River and dealt with their envy and accusations constantly." Edwards added, "But as the joke goes: '. . . ach, you fuck one sheep!'"

Following Edwards's resignation, relations between the boat shop and vegetation devolved into a cold war. On trips, according to multiple sources, some of the boatmen withheld food or avoided taking volunteers to work sites. Watters complained to the director of the science division and to regional Park Service authorities. After getting nowhere, she quit in frustration, and Anne eventually assumed her place. According to Anne and Hall, Loeffler later showed up at a campsite where Anne was working to harangue her about Edwards. He and Desrosiers made it so difficult for her to schedule trips that sometimes she had to use a helicopter, at great expense. These acts of sabotage "became an art form for the two Daves," recalled Hall. The pair even erected a memorial to Edwards in the boat shop, said two former employees: a crude bust of Jesus wearing a crown of thorns with Edwards's name written on the base in Sharpie. The implication was clear: Edwards had been martyred.

It was around the time of Edwards's departure that Szydlo took her boat trip with Loeffler. After the scare in the rapids, she said, the uneasy balance between them shifted. Szydlo stopped laughing at his come-ons. Loeffler would sleep in late and then tell her they didn't have time to visit her next work site. "This person was in complete control of everything I needed to survive," she said. "I was terrified." She began to formulate a plan to get out of the canyon if she needed to. "Even if there were trails to take, which in most places there were not, they'd land me in the middle of nowhere, in the desert, up on the rim," she said. "I didn't have enough food or water to attempt that." She could try to hike out on the Bright Angel Trail when they reached the halfway point at Phantom Ranch. But doing so would mean missing the nesting sites on the lower half of the river—and, she feared, abandoning any hope of being hired back next season.

The day before they reached Phantom Ranch, Szydlo said she felt as if some kind of assault was inevitable. Loeffler slowed the motorboat to a crawl, stopping at nearly every beach. Finally, in the middle of a channel, she heard the motor go quiet. Loeffler came up behind her, grabbed her shoulders, and asked her to describe her sexual fantasies so he could act them out.

"I broke down crying," Szydlo said. "Saying, 'Get off me, stop harassing me.' As soon as I used the word *harassment,* he was like, 'Whoa, stop. I don't know what you're talking about.'" He revved the engine and sped to Phantom Ranch. For the last five days, she said, they barely spoke, and at meals Loeffler gave her minuscule portions. After she returned, she emailed her then-boyfriend and told him what had happened. Szydlo worried for months about whether she should file a report. When she finally contacted an HR representative almost six months later, she said, she received a brief response informing her she'd need dates, times, and witnesses in order to pursue a complaint. She let it drop, not wanting to start a "huge, ugly fight." Much as she suspected, other women in similar situations have discovered that taking formal action can bring on its own host of problems.

The Eldorado National Forest is a mountainous expanse of nearly 1,000 square miles that stretches from east of Sacramento to the crest of the Sierra Nevada. Denice Rice has worked here for 15 years as a firefighter—on engines and fire crews and as a prevention officer. These days she likes to operate by herself, driving a truck with a small reserve of water through the hundreds of miles of back roads that cut into the Eldorado. She is often the first on the scene at a fire, helping direct in crews of "hot shots," the firefighting elite who clear the tree line. On slower days she might serve as "Smokey's wrangler," accompanying the unlucky staffer who has to don the sweaty mascot costume and make safety presentations to kids.

Many women in the Forest Service told me that "fire is a small world" and that they repeatedly had to fight the perception that they were only there to meet men. Rice, who exudes a no-bullshit air of competence, prided herself on her toughness. When I visited her at her home in January, she drove to meet me on a four-wheeler, flanked by two bulldogs. "When you work in fire, you have to have a really thick skin," she said.

Around 2008, Rice was a captain being groomed for promotion when she was befriended by her boss's boss, a division chief named Mike Beckett. After about a year their interactions took on a different tone. By Rice's account, Beckett would describe sexual dreams he'd had about her and comment on her body. When they texted about work, he responded with crass double-entendres. He cornered her in the office, followed her into the bathroom, and tried to touch her or lift her shirt. She said he groped or touched her inappropriately at least 20 times.

Even when she was out in the field, Rice felt as if there was no escape. Sometimes Beckett would wait late for her to return to the office. He took to radioing in to ask her location and seemed to monitor the line for word of her whereabouts: he'd appear, unannounced, when she was in some remote location—say, a tower lookout high in the Sierras. "He was paying a lot of attention to an employee three to four pay grades below him, which is uncommon," recalled Rice's former direct supervisor, who still works at the Forest Service. "He was constantly going around me."

It became so uncomfortable that Rice stopped calling in her location—a significant safety risk. Eventually Beckett arranged for her to be moved out of the office she shared with a colleague and into a room on her own. It was more of a storage area, recalled the former supervisor, tucked in the back of the building. During this time her oversight duties were stripped from her one by one, Rice later said in a signed affidavit, and the former supervisor confirmed in an interview. (Beckett declined to answer any questions, and the Forest Service said it couldn't comment on specific allegations.)

Still, Rice was reluctant to take formal action. She didn't want to be "one of those women," she explained. "You don't cry in front of the guys, you don't show weakness in front of them. And you don't file. You just don't file. You suck up and deal." But one day in 2011, she said, after three years of harassment, Beckett came into her office and with a letter opener poked her repeatedly on her chest, drawing a circle around her nipple. She filed. Randy Meyer, the Eldorado union steward, said he got a phone call from Rice "that scared me to death. She was highly emotional and beside herself." He told a senior forest manager that he was prepared to alert the police—and "then everybody and his brother got involved in this mess."

In the ensuing investigation, some 30 of Rice's and Beckett's colleagues were interviewed about humiliating details that Rice hadn't even confided to her husband. "Everybody knew that he took me in the bathroom, tried to take my clothes off, things that he would say to me: 'I want to watch you pee.' They all knew," she said. "And I still work with these people." Rice said she got sick from the stress. The supervisor added that once, after he went to check on Rice, Beckett threatened him with disciplinary action.

In 2012, at the district ranger's request, Rice's supervisor called an all-hands meeting. Rice was certain that Beckett would be on the agenda. She begged not to have to attend but said she was required to show up. (Rice's former supervisor couldn't verify this but said the meeting was handled insensitively: "Nobody took into consideration that maybe she was still feeling like the target in the case.") The situation with Beckett was discussed in front of at least 50 colleagues; Rice walked out in tears. "I think that was the worst thing that ever happened to me," she said.

When we spoke, Rice was jumpy and broke down several times. "I can't go anywhere without wondering, 'Do people know who I am?'" she said. One male firefighter who has worked with Rice for five years told me, "It changed her whole life. People know Denice's story on the forest, so she has this cloud around her. I've seen it for four years. I see Denice 'trigger' all the time: in classroom settings, out in the woods."

Ultimately the ranger in charge of the investigation recommended that Beckett should be fired. But Beckett retired before any action could be taken. Meanwhile Rice's career has effectively stalled. The firefighter who worked with Rice requested anonymity, explaining, "If the powers that be tie me to her in any way, I'll never promote here again."

Rice's ordeal wasn't unique. Lesa Donnelly said that in her capacity as an advocate, she has been contacted by scores of women in the service in California who allege they've been punished for pursuing sexual harassment complaints. One 22-year-old forestry technician filed a claim and several days afterward was visited by officials who searched only her side of the barracks with a drug dog. According to a subsequent complaint she lodged with the Forest Service, her roommate told her that one official had remarked, "You guys must have pissed someone off." The woman left the service soon afterward.

Elisa Lopez-Crowder, a 34-year-old navy veteran, was hired as a firefighter in 2010. She ran 45-pound sections of hose into the forest and cleared live trees to create fuel breaks. In her first months on the Eldorado, she said, an assistant captain asked her whether she'd been a "bitch" or a "slut" in the navy, and whether her skin was really that color or just dirty. One day while she was clearing brush, she claimed, he hoisted her by her line gear and threw her to the ground; according to a male coworker's account, he held her down with his foot. The coworker intervened, and later joined her to report the matter to their captain.

The assistant captain was briefly placed on administrative leave. (In a court declaration he said Lopez-Crowder had "tripped" and that "before I helped her up, I jokingly placed my foot on her pack.") While an investigation was still underway, he was assigned to the same work sites as Lopez-Crowder. About a year later she traveled with Donnelly and other Forest Service women to bring their concerns to USDA Secretary Tom Vilsack in Washington, D.C. Lopez-Crowder said Vilsack apologized and assured her that the assistant captain had been removed from his position; it fell to Lopez-Crowder to tell the secretary that he was still on the Forest Service payroll. A short time later the assistant captain left the force. Lopez-Crowder transferred out of the firefighting division anyway, fearing that she had become a target. "In the years I served in the military," she said, "I never encountered such discrimination and harassment as I have working for the U.S. Forest Service."

Alicia Dabney, a mother of three who lives on the Tule River Indian reservation, became a firefighter, like her father and uncles before her, at the age of 26 in the Sequoia National Forest. According to Equal Employment Opportunity complaints she filed in 2011 and 2012, Dabney claimed that coworkers made disparaging remarks about her Latina and Comanche heritage and joked about sexually assaulting women. She said a male supervisor instructed her and another female firefighter to tell him when they began menstruating. At a training academy, other participants left lewd sexual propositions on her voicemail. One day she arrived at work to find the floor of the engine house strewn with printouts that read "Alicia Dabney The Whore." (She provided a photo of the printouts.)

Some of the harassment was physical. Once a male coworker jumped on her neck, "riding me like a big horse," she recalled. On an assignment in Texas, she said, a supervisor put her in a chokehold and threw her on his hotel bed. A USDA investigation substantiated the first of those incidents but denied that there had been a "pattern of harassment." In 2012, Dabney was informed that the Forest Service was initiating her termination, claiming she had omitted part of her criminal record—a misdemeanor vandalism charge—and failed to disclose federal debt on her application. (Dabney maintains that she disclosed both.) In 2013, Dabney left and signed a settlement agreement with the Forest Service.

In 2011 the USDA put the Forest Service into temporary receivership for its failure to adequately respond to sexual harassment claims. For the next year all EEO complaints were handled by the secretary's office in Washington. Tom Tidwell, the chief of the Forest Service, explained in an email to staff that the change would allow the agency "to better process a series of EEO complaints within the Forest Service that, frankly, we have not handled well."

In the Canyon's River District, the problems had continued unabated since Cheyenne Szydlo's 2006 trip. Certain boatmen were repeatedly accused of harassing or assaulting women in strikingly similar scenarios. One young boatman covered his Park Service boat hatch with pictures of topless women and boasted to coworkers, including Dan Hall, about a side gig recruiting college women for Girls Gone Wild–style videos. Hall said that half a dozen intermittent boatmen who, like him, objected to the boat shop's culture found themselves blacklisted from river assignments. And even in the rare cases when management did take swift action, the targets weren't always the people you'd expect.

In 2011, Mike Harris, a contract hire then in his late 50s, was training a 40-year-old river ranger named Chelly Kearney to operate a new boat. She said that he directed her to pull to the shore, away from their group, and announced that he was going to take a bath. Then, she said, he removed all of his clothes and invited Kearney to join him in the water. When Kearney asked if they could leave, he put on his life jacket and climbed back on the boat naked. He "stood there with his penis completely exposed," Kearney later wrote in a detailed letter to park leadership. "I stated to Harris, 'Do not get on this boat until you put your clothes on.' He stated to me that he needed to dry his clothes out. I said, 'No, do not get on this boat without your clothes.' He finally put on a pair of long underwear pants." Harris confirmed to me that he climbed onto the front of the boat naked: "I just wanted to sit in the sun and dry out," he said. However, he said he thought he had permission from Kearney to bathe and didn't ask her to join him.

Upon Kearney's return, she said she told a supervisor about the incident. The supervisor, she alleged, joked that they "used to not call it sexual harassment until the guy whipped out his penis and slapped you across the face with it." Kearney didn't take the matter further.

The next year, on another trip, a biologist I'll call Lynn said Harris repeatedly asked her to sleep in his tent when hers started leaking during a rainstorm. After she refused, he set up his tent directly next to hers. Harris told me that he only asked Lynn to join him in his tent once and hadn't meant the invitation as a come-on. "It wasn't to have sex," Harris said. "I think I said something like 'We could snuggle and that's all.'"

Lynn said she emailed her supervisor about the episode. After a third female employee filed an EEO complaint about his behavior in 2013, Harris resigned. Lynn's complaint was supposed to be confidential, but she noticed that boatmen she'd been friendly with began to act coldly toward her. And matters only escalated from there.

In February 2014, Dave Loeffler led a joint Park Service–private sector trip. Both Anne and Lynn were apprehensive about being on the river with him. At one point, Lynn said, a passenger inquired about a boatman who'd been let go and Loeffler ranted about "complainers" who had ruined boatmen's lives. The following day, as the group approached a campsite, Lynn was standing in the bow of her boat when Loeffler pulled her out roughly by her life jacket—a shocking breach of river norms. Anne came up to Lynn on the beach to find her concealing tears behind her sunglasses. Lynn wanted to leave, but at that point there was no way for her to hike out.

On the last night the party celebrated with dinner and drinks. A woman who worked for a private boat company produced a novelty penis-shaped straw she'd received at a bachelorette party and dropped it in a colleague's drink. People laughed and passed the straw around. At one point Lynn was holding it when Loeffler tried to take her picture. Then someone put on music. It was an eclectic playlist, and people danced accordingly: interpretive dance, head-banging, two-stepping. A hip-hop song came on, and the group started talking about twerking. Lynn gave a comically awkward demonstration in her heavy canvas Carhartt pants, puffy down jacket, and rubber boots.

Two days later Anne and Lynn were called into the offices of upper management and informed that they'd been accused of sexual misconduct. In written statements, Loeffler and two of his friends claimed that Anne and Lynn had shoved the penis straw in Loeffler's face, danced provocatively in short skirts, and, as one complainant put it, behaved "coquettishly" throughout the trip. "I felt I needed to remove myself from this increasingly hostile work environment," Loeffler wrote in his statement. "They were being so rude and inappropriate to myself and others." According-ing to notes from the manager assigned to look into the situation, Loeffler said he wanted Anne and Lynn to be "treated similarly" to other employees accused of harassment—that is, with the Park Service deciding not to renew their contracts.

Both women protested to the managers that they were being retaliated against for their previous reports of sexual harassment. Nonetheless, the park launched an investigation, although both superintendent David Uberuaga and deputy superintendent Diane Chalfant would later acknowledge in an official report that it may not have been thorough enough. In particular, the investigators weren't made aware of the history between Anne, Lynn, and the boatmen.

In a meeting Lynn said Chalfant told her that Loeffler's charges couldn't be retaliatory, since Lynn's previous sexual harassment complaint was confidential. Both Lynn and Anne were informed that their contracts would not be renewed. In Lynn's termination letter, Chalfant wrote, "We cannot afford to have team members in our employment who are not on board with management's expectations and requirements."

"What happened to [Lynn] was the most horrifying thing I'd ever seen," said Chelly Kearney, who had made her own efforts to draw attention to the treatment of women on the river. About a year after she resigned in 2012, she wrote a 29-page letter to Grand Canyon chief ranger Bill Wright documenting multiple instances of harassment, assault, and retaliation and describing a culture that protected male harassers while allowing victims to be targeted for retaliation. The Park Service requested a formal EEO investigation, but the final report was never distributed beyond the uppermost level of park management and no disciplinary actions were taken.

Following Lynn's and Anne's dismissal, Kearney tried again. She forwarded her letter to Uberuaga, writing that she had witnessed a "disturbing and pervasive level of hatred" toward Anne and her boss and that Anne should be protected by federal whistle-blower laws. She received a brief response from Uberuaga thanking her for her concern.

Some former park employees now ruefully refer to the fateful party as "The Night on Cock-Straw Beach," and the incident became an unlikely rallying point. Hall sent around an email asking a core group of former park employees and colleagues in private rafting companies to gather names of other women who'd been harassed or run out of the River District. With Donnelly's help, 12 women and Hall wrote to secretary of the interior Sally Jewell, requesting a formal investigation into the "pervasive culture of discrimination, retaliation, and a sexually hostile work environment" in the River District.

Where Donnelly had tried for decades to get federal authorities to intervene more decisively in the Forest Service, the DOI responded quickly. In October its Office of Inspector General launched an investigation that grew from the 13 initial complainants to include multiple interviews with more than 80 people. Their final report would identify 22 additional victims or witnesses. It included accounts of Cheyenne Szydlo's 2006 trip with Loeffler, the Halloween party where Edwards took the photo up Anne's skirt, the twerking incident that led to the complaint against Anne and Lynn, and several allegations involving a boatman that a former employee identified as Mike Harris.

The women's complaints, the investigators said, were "extremely credible." The investigators also determined that Chalfant, the deputy superintendent, had allowed the complaint letter signed by the 12 women and Hall to make its way to some of the accused boatmen, in violation of policy. In an interview the lead investigator, Greg Gransback, criticized the park's handling of the accusations against Anne and Lynn. "If you compare what had happened to these two in the past and what they were accused of, I mean there's just no comparison. It's apples and oranges," he said. "The park got it wrong where they went overboard."

In a February response to the investigation, the Park Service's Intermountain Region didn't contest any of the details in the report and admitted that in many instances appropriate action hadn't been taken. In the OIG report two boatmen whose actions are clearly consistent with those of Loeffler and Desrosiers deny all allegations made against them. (I was unable to reach Desrosiers directly despite contacting the Park Service, former colleagues, and two family members.) Boatman 3—whom a former employee identified as Loeffler—told the OIG that he "acknowledged making sexual remarks to women, but said that he did so only when he sensed a 'mutual attraction.'" James Doyle, the communications chief for the Intermountain Region, said he couldn't discuss individual allegations against employees and added, "We maintain a zero tolerance for sexual harassment and hostile workplace environment."

During the year and a half that the investigation was underway, the park made some changes. After Bill Wright transferred out of the district, his role was filled by a woman. The policy for staff boat trips was revised. There would be no alcohol permitted and an outside supervisor would be required on all expeditions. Dave Desrosiers retired in May 2015. According to its response to the OIG, the park is introducing a detailed plan to improve its sexual harassment policies and considering disciplinary action against managers who mishandled complaints. All employees are now required to wear "standard uniforms" on river trips.

The OIG team was more than familiar with sexual harassment cases: Gransback had worked on the inquiry that resulted from the 1996 Aberdeen Proving Ground scandal, when 12 army officers were charged with assaulting female trainees. Still, Gransback told me that even he and his seasoned colleagues teared up when they heard Grand Canyon women describe the fine line they had to walk to do their jobs, "between not being hated and not being desired."

In the Tailhook case, he noted, the accused military members had developed a *Top Gun* mentality, believing they were too important to be taken down. He observed the same dynamic at work among the boatmen. "They became almost untouchable," he said. But the military, Gransback pointed out, has made "drastic changes," including evidence-based sexual harassment and assault prevention programs. So far neither the Park nor the Forest Service has proposed anything so extensive. (Since June 2015 the Forest Service's California region has strengthened its protocols for sexual harassment training and reporting, a spokesperson said.)

In my conversations with the women, they expressed great pride in their strength. For years they had performed dangerous, physically demanding jobs. Many of them had faced life-threatening situations. All of them had operated within environments in which women had very little room for error. The harassment they described had not only brought about personal humiliation or the loss of a job or even a career. It had shaken their entire perception of themselves—as tough and resilient, able to handle anything that man or nature could throw at them.

They lost other things too. After her boat trip with Loeffler, Cheyenne Syzdlo found herself avoiding the river. "When I'd hear people talk about how much they loved river trips, I'd be like, 'Oh God, I hated them, I hated them,'" she told me. Then, in the course of our conversations, she came across an email she'd written to a friend after her second time in the Grand Canyon, before she'd ever met Dave Loeffler.

In her message Syzdlo described the thrill of riding huge rapids in the bow of an inflatable boat. She remembered how even the most experienced guides would pause and become tense, studying the water before steering them in. She recalled the night her group camped on a sliver of beach when a thunderstorm suddenly erupted, sending loose boulders tumbling down the sheer cliff face. She and her colleagues had huddled in their tents and contemplated the possibility that they might die, and then, when the morning dawned damp and bright, laughed as they fished their supplies out of the river. "I'd never thought about that second trip again because the third trip did change everything. It was magical," she told me. "It's so primitive and you feel so free. You never experience that in life." She'd forgotten about it for nearly a decade, but that morning on the river, she hadn't wanted to leave.

JON MOOALLEM

The Amateur Cloud Society That (Sort of) Rattled the Scientific Community

FROM *The New York Times Magazine*

GAVIN PRETOR-PINNEY DECIDED to take a sabbatical. It was the summer of 2003, and for the last 10 years, as a sideline to his graphic-design business in London, he and a friend had been running a magazine called *The Idler*. *The Idler* was devoted to the "literature for loafers." It argued against busyness and careerism and for the ineffable value of aimlessness, of letting the imagination quietly coast. Pretor-Pinney anticipated all the jokes: that he'd burned out running a magazine devoted to doing nothing, and so on. But it was true. Getting the magazine out was taxing, and after a decade it seemed appropriate to stop for a while and live without a plan—to be an idler himself and shake free space for fresh ideas. So he swapped his flat in London for one in Rome, where everything would be new and anything could happen.

Pretor-Pinney is 47, towering and warm, with a sandy beard and pale blue eyes. His face is often totally lit up, as if he's being told a story and can feel some terrific surprise coming. He stayed in Rome for seven months and loved it, especially all the religious art. One thing he noticed: the paintings and frescoes he encountered were crowded with clouds. They were everywhere, he told me recently, "these voluptuous clouds, like the sofas of the saints." But outside, when Pretor-Pinney looked up, the real Roman sky was usually devoid of clouds. He wasn't accustomed to such endless blue emptiness. He was an Englishman; he was accustomed to clouds. He remembered as a child being enchanted by them and deciding that people must climb long ladders to harvest cotton from them. Now, in Rome, he couldn't stop thinking about clouds. "I found myself missing them," he told me.

Clouds. It was a bizarre preoccupation, perhaps even a frivolous one, but he didn't resist it. He went with it, as he often does, despite not having a specific goal or even a general direction in mind; he likes to see where things go. When Pretor-Pinney returned to London, he talked about clouds constantly. He walked around admiring them, learned their scientific names and the meteorological conditions that shape them, and argued with friends who complained they were oppressive or drab. He was realizing, as he later put it, that "clouds are not something to moan about. They are, in fact, the most dynamic, evocative, and poetic aspect of nature."

Slowing down to appreciate clouds enriched his life and sharpened his ability to appreciate other pockets of beauty hiding in plain sight. At the same time, Pretor-Pinney couldn't help noting, we were entering an era in which miraculousness was losing its meaning. Novel, purportedly amazing things ricocheted around the Internet so quickly that, as he put it, we can now all walk around with an attitude like, "Well, I've just seen a panda doing something unusual online, what's going to amaze me now?" His fascination with clouds was teaching him that "it's much better for our souls to realize we can be amazed and delighted by what's around us."

At the end of 2004 a friend invited Pretor-Pinney to give a talk about clouds at a small literary festival in Cornwall. The previous year there were more speakers than attendees, so Pretor-Pinney wanted an alluring title for his talk, to draw a crowd. "Wouldn't it be funny," he thought, "to have a society that defends clouds against the bad rap they get—that stands up for clouds?" So he called it "The Inaugural Lecture of the Cloud Appreciation Society." And it worked. Standing room only! Afterward people came up to him and asked for more information about the Cloud Appreciation Society. They wanted to *join* the society. "And I had to tell them, well, I haven't really got a society," Pretor-Pinney said.

He set up a website. It was simple. There was a gallery for posting photographs of clouds, a membership form, and a florid manifesto. ("We believe that clouds are unjustly maligned and that life would be immeasurably poorer without them," it began.) Pretor-Pinney wasn't offering members of his new Cloud Appreciation Society any perks or activities, but to keep it all from feeling ephemeral or imaginary, as many things on the Internet do, he eventually decided that membership should cost $15 and that members would receive a badge and certificate in the mail. He recognized that joining an online Cloud Appreciation Society that only nominally existed might appear ridiculous, but it was important to him that it not feel meaningless.

Within a couple of months the society had 2,000 paying members. Pretor-Pinney was surprised and ecstatic. Then Yahoo placed the Cloud Appreciation Society first on its 2005 list of Britain's "Weird and Wonderful websites." People kept clicking on that clickbait, which wasn't necessarily surprising, but thousands of them also clicked through to Pretor-Pinney's own website, then paid for memberships. Other news sites noticed. They did their own articles about the Cloud Appreciation Society, and people followed the links in those articles too. Previously Pretor-Pinney proposed writing a book about clouds and was rejected by 28 editors. Now he was a viral sensation with a vibrant online constituency; he got a deal to write a book about clouds.

The writing process was agonizing. On top of not actually being a writer, he was a brutal perfectionist. But *The Cloudspotter's Guide,* published in 2006, was full of glee and wonder. Pretor-Pinney relays, for example, the story of the United States Marine pilot who in 1959 ejected from his fighter jet over Virginia and during the 40 minutes it took him to reach the ground was blown up and down through a cumulonimbus cloud about as high as Mount Everest. He surveys clouds in art history and Romantic poetry and compares one exceptionally majestic formation in Australia to "Cher in the brass armor bikini and gold Viking helmet outfit she wore on the sleeve of her 1979 album *Take Me Home.*" In the middle of the book there's a cloud quiz. Question No. 5 asks of a particular photograph, "What is it that's so pleasing about this layer of stratocumulus?" The answer Pretor-Pinney supplies is, "It is pleasing for whatever reason you find it to be."

The book became a bestseller. There were more write-ups, more clicks, more Cloud Appreciation Society members. And that cycle would keep repeating, sporadically, for years, whenever an editor or blogger happened to discover the society and set it off again. (There are now more than 40,000 paid members.) The media tended to present it as one more amusing curiosity, worth delighting over and sharing before moving on. That is, Pretor-Pinney's organization was being tossed like a pebble, again and again, into the same bottomless pool of interchangeable online content that he was trying to coax people away from by lifting their gaze skyward. But that was okay with him; he understood that it's just how the Internet works. He wasn't cynical about it, and he didn't feel his message was being cheapened either. It felt as if he were observing the whole thing from afar, and he tried to appreciate it.

Then Pretor-Pinney noticed something odd.

"The way I felt when I first saw it was: Armageddon," Jane Wiggins said. Wiggins was a paralegal working in downtown Cedar Rapids, Iowa, in June 2006 when she looked out her office window and saw an impenetrable shroud of dark clouds looming over town. Everyone in the office stood up, Wiggins told me, and some drifted to the window. The cloud was so enormous, so terrible and strange, that it made the evening news. Wiggins, who had recently taken up photography, took out her camera.

Soon after that Wiggins discovered the Cloud Appreciation Society website and posted one of her pictures in its gallery. But the anomaly Wiggins thought she had captured wasn't actually anomalous. Similar photos turned up in the Cloud Appreciation Society's gallery from Texas, Norway, Ontario, Scotland, France, and Massachusetts. Pretor-Pinney assumed that this phenomenon was so rare that until now no one had recognized it as a repeating form and given it a name. "As the hub of this network, a network of people who are sky-aware," he said, "it's easier to spot patterns that perhaps weren't so easy to spot in the past."

In fact, many aspects of meteorology already rely on a global network of individual weather observers to identify cloud types with the naked eye, filing them into a long-established scientific framework: not just as cumulus, cirrus, stratus, or cumulonimbus clouds, as schoolchildren learn, but within a recondite system for describing variations. Atypical clouds are either fitted into that existing map of the sky or set aside as irrelevant. Pretor-Pinney liked classifying clouds using these names; he was thankful to have that structure in place. And yet it seemed a shame to repress the glaring, deviant beauty recorded in Wiggins's photograph by assigning it a name that didn't sufficiently describe it. He supposed, if you had to, you could call this thing an undulatus—the standard classification for a broad, wavy cloud. But that seemed to be selling the cloud tragically short, stubbornly ignoring what made it so sublime. This was "undulatus turned up to 11," he said. So he came up with his own name for the cloud: asperatus. (The word *asperatus* came from a passage in Virgil describing a roughened sea; Pretor-Pinney had asked his cousin, a high school Latin teacher, for help.) He wondered how to go about making such a name official.

In 2008, while shooting a documentary for the BBC about clouds, Pretor-Pinney pitched his new cloud to a panel of four meteorologists at the Royal Meteorological Society. The scientists sat in a line behind a table; Pretor-Pinney stood, holding blown-up photos of asperatus for them to consider. "It was a lot like *The X Factor,*" he said, referring to the TV talent show. The scientists were encouraging but diplomatic. A new cloud name, they explained, could be designated only by the World Meteorological Organization, an agency within the United Nations, based in Geneva, which has published scientific names and descriptions of all known cloud types in its *International Cloud Atlas* since 1896. The WMO is exceptionally discerning; for starters, Pretor-Pinney was told, he would need more carefully cataloged incidences of these clouds, as well as a scientific understanding of their surrounding "synoptic situation." The process would take years. And even then, the chances of inclusion in the atlas were slim. The WMO hadn't added a new cloud type to the *International Cloud Atlas* since 1953. "We don't expect to see new cloud types popping up every week," a WMO official named Roger Atkinson told me. When I asked why, Atkinson said, "Because 50 or 60 years ago we got it right."

A cloud is only water, but arranged like no other water on earth. Billions of minuscule droplets are packed into every cubic foot of cloud, throwing reflected light off their disordered surfaces in all directions, collectively making the cloud opaque. In a way, each cloud is an illusion, a conspiracy of liquid masquerading as a floating, solid object.

But for most of human history, what a cloud was, physically, hardly mattered; instead we understood clouds as psychic refuges from the mundane, grist for our imaginations, feelings fodder.

Clouds both influenced our emotions and hung above us like washed-out mirrors, reflecting them. The English painter John Constable called the sky the "chief organ of sentiment" in his landscapes. And our instinct as children to recognize shapes in the clouds is arguably one early spark of all the higher forms of creative thinking that make us human and make us fun. Frankly, a person too dull to look up at the sky and see a parade of tortoises or a huge pair of mittens or a ghost holding a samurai sword is not a person worth lying in a meadow with. In *Hamlet,* Polonius's despicable spinelessness is never clearer than when Hamlet gets him to enthusiastically agree that a particular cloud looks like a camel, then not a camel at all but a weasel. Then not a weasel but a whale. Polonius will see whatever Hamlet wants him to; he is a man completely without his own vision.

We look for meaning—portents—in the clouds as well, the more grown-up version of picking out puffy animals. "There's a long history of people finding signs in the sky," Pretor-Pinney told me, from Constantine seeing the cross over the Milvian Bridge to the often belligerent protesters outside Pretor-Pinney's talks, who are convinced that the contrails behind commercial airplanes are evidence of a toxic, secret government scheme and are outraged that Pretor-Pinney—the righteous Lorax of clouds—refuses to expose it. In short, clouds exist in a realm where the physical and metaphysical touch. "We look up for answers," Pretor-Pinney says. And yet we often don't want empirical answers. There has always been a romantic impulse to protect clouds from our own stubbornly rational intellects, to keep knowledge from trampling their magic. Thoreau preferred to understand clouds as something that "stirs my blood, makes my thought flow" and not as a mass of water. "What sort of science," he wrote, "is that which enriches the understanding but robs the imagination?"

The scientific study of clouds grew out of a collection of madly appreciating amateurs who struggled with this same tension. The field's foundational treatise was first presented to a small scientific debating society in London one evening in 1802 by a shy Quaker pharmacist named Luke Howard. Howard, then 30, was not a professional meteorologist but a devoted cloud-spotter with a perceptive, if wandering, mind. His interest in clouds started early. His biographer, Richard Hamblyn, explains that as a young student in Oxfordshire, Howard seems to have found school magnificently boring. He couldn't bring himself to pay attention, except to his Latin teacher, who punished daydreaming with beatings. Today Howard might covertly pull out his phone and read a link a friend shared about, say, an eccentric society in England that appreciates clouds. But poor Howard's boredom was analog: all he could do was look out the classroom window at the actual clouds rolling by.

Howard's intention that night in London was to bring clouds down to earth without depleting their loftiness. After years of closely observing clouds, his appreciation of them had hardened into analysis. He now insisted that though clouds may appear to be blown around in random, ever-changing shapes, they actually take consistent forms, forms that can be distinguished from one another and whose changes correspond to changes in the atmosphere. Clouds can be used to read what Howard called "the countenance of the sky"; they are an expression of its moods, not just in a poetic way, as Constable meant, but meteorologically.

Howard's lecture was eventually published as "On the Modifications of Clouds, and on the Principles of Their Production, Suspension and Destruction." It stands as the ur-text of nephology, the branch of meteorology devoted to clouds. Howard divided clouds into three major types and many intermittent varieties of each, all similarly affixed with Latin names or compounds. (He had learned his Latin well.) Like Linnaeus, who used Latin to sort the fluidity of life into genera and species, Howard used his new cloud taxonomy to wrest our understanding of the world's diversity from superstition and religion. His signature assertion that "the sky, too, belongs to the Landscape" can be read as a call for empiricism—a conviction that science can in fact measure out the mystical.

Nearly a century later Howard's work would be picked up by another energetic amateur, the Honorable Ralph Abercromby. Abercromby was the bookish great-grandson of a celebrated English war hero. He was apparently so meek and frail ("never robust, even as a boy," one tribute read after his death) that he was forced to drop out of school and was rarely able to hold a job. He served briefly in the military but seemed completely unsuited to soldiering; deployed to Newfoundland in 1864, Abercromby began theorizing about how the fog there was produced. Later, stationed in Montreal, he scrutinized the wind. It would have been tempting for his superiors to label him "absent-minded" or "unfocused," but in retrospect it was just another case of a young man intensely focused on something few people considered worthy of attention —another case of a young man in love with clouds.

In 1885, Abercromby took his first round-the-world voyage. He was a civilian again, and his private physician hoped the sea air would restore his pitiable health. But he worked slavishly the whole time, keeping a meticulous weather diary, photographing the clouds at sea. He published many scientific papers and a book about the clouds and weather that he encountered. And he kept traveling: Scandinavia and Russia, Asia and the United States, compelled, as he wrote, to "continue the observation and photography of cloud forms in different countries." Looking up, Abercromby came to realize that clouds looked essentially the same everywhere. Colonialism was sending goods, resources, and culture around the planet; suddenly it must have seemed obvious that we also shared the same sky.

Abercromby's primary interest was in refining the science of weather-tracking and forecasting, and he knew that meteorologists everywhere would need a standard way to discuss and share their observations. Eventually, collaborating with a Swedish cloud scientist named Hugo Hildebrand Hildebrandsson, he convened a Cloud Committee to hammer out Hildebrandsson's meticulous "Nomenclature of Clouds." They declared 1896 "the International Year of the Cloud." By year's end the committee produced the first *International Cloud Atlas.*

The atlas is now in its seventh edition, and its meticulous taxonomy provides for 10 genera of clouds, 14 species, nine varieties, and dozens of "accessory clouds" and "supplementary features." The atlas also establishes a grammar with which these terms can be combined to allow for the instability of clouds—the way they morph from one form into another—or to describe their general altitude. A cumulus, for example, might just be a cumulus; or it might be a cumulus fractus, if its edges are tattered; or a cumulus pileus, if a smaller cloud appears over it like a hood. An altocumulus lenticularis, meanwhile, is a vast, tightly bunched flock of clouds stretching across the sky at altitudes from 6,500 to 23,000 feet.

Of course, not everything in the sky needs to be precisely described. As a reference book for meteorologists, the atlas has been concerned only with clouds that have "operational significance"

—that reliably reveal something about atmospheric conditions. As far as other clouds go, says Roger Atkinson of the WMO, one person might look at a cloud and say, "'It's wonderful. It looks like an elephant,' and someone else might think it's a camel." But the WMO doesn't particularly care. It does not see its mission as settling disagreements about elephants and camels.

Soon after Pretor-Pinney appeared on the BBC, championing his asperatus cloud, the media seized on the possibility, however remote, that the WMO would add asperatus to its atlas. Suddenly there were stories about the Cloud Appreciation Society all over the place, all over again. This time Pretor-Pinney—previously cast as a charming English eccentric with a funny website—was presented as the crusading figurehead of a populist meteorological revolt. Pretor-Pinney had initially turned defeatist after shooting the documentary and never bothered reaching out to the WMO; the bureaucracy seemed too formidable. Now he didn't quite know what to say. When reporters called, he suggested they contact the WMO, impishly channeling them as de facto lobbyists.

Then, in 2014, the WMO announced it was preparing the first new edition of the *Cloud Atlas* in nearly 40 years; the agency felt pressure to finally digitize the book, to reassert its authority over the many reckless cloud-reference materials proliferating online. One of the WMO's first steps was to convene an international task team to consider additions to the atlas. "Most public interest," a news release noted, "has focused on a proposal by the Cloud Appreciation Society" to recognize the so-called asperatus. The task team would report to a so-called Commission for Instruments and Methods of Observation. Last summer the commission recommended to the World Meteorological Organization's 17th World Meteorological Congress in Geneva that the cloud be included. Everyone seemed confident that the recommendation would soon be ratified by the WMO's executive council. Except the new cloud wasn't asperatus anymore; it was now asperitas. The task team had demoted it from a cloud "variety," as Pretor-Pinney had proposed, to a "supplementary feature," and the elaborate naming convention for clouds required supplementary features to be named with Latin nouns, not adjectives. "One of those things that's so close, but different," Pretor-Pinney told me, with a tinge of amusement and resentment.

When I spoke to Roger Atkinson, of the WMO, he stressed that asperitas would merely be "a fourth-order classification, not a primary genus, not one of the primary cloud types, not one of the Big Nine." Neither was it the only new classification the task team recommended adding; it was just the most famous one. The prominence of the cloud seems to have forced the scientists' hand. Asperitas didn't appear to have any operational significance, but the public enthusiasm Pretor-Pinney had gathered around the cloud ultimately made asperitas too prominent to ignore. One task-team member, George Anderson, told me that not giving such a well-known cloud a definitive name would only create more confusion.

Pretor-Pinney conceded all this, happily. "My argument is not that this is some hugely significant thing," he told me. By now he was mostly using the cloud to make a point—to needle the "human vanity" inherent in "the Victorian urge to classify things, to put them into pigeonholes and give them scientific names." Clouds, he added, "are ephemeral, ever-changing, phenomenal. Here you have a discrete, scientific, analytic urge laid onto the embodiment of chaos, onto these formations within these unbounded pockets of our atmosphere where there's no beginning and no edge." All he wanted was to encourage people to look at the sky, to elevate our perception of clouds as beautiful "for their own sake."

Slowly, over the last 200 years, the impulse of cloud lovers like Howard and Abercromby to make the mystical empirical had ossified into something stringent and reductive. Pretor-Pinney wanted to clear a little more space in our collective cloudscape for less distinct feelings of delight and wonder. His championing of asperatus was in reality somewhat arbitrary. There were a few other unnamed cloud forms he saw repeating in the society's photo gallery. He just happened to pick this one.

The cultural history of clouds seemed to be shaped by a procession of amateurs, each of whom projected the ethos of his particular era onto those billowing blank slates in the troposphere. Pretor-Pinney was our era's, I realized—the Internet era's. He wasn't just challenging the cloud authorities with his crowdsourced cloud; he was trolling them.

I was one of the many reporters who contacted Pretor-Pinney when the first photos of asperitas made the rounds in 2009. I had seen an Associated Press article, with Jane Wiggins's photo of the cloud

in Iowa and a reference to Pretor-Pinney and his Cloud Appreciation Society, and felt a kind of instant and exhilarated envy: apparently some people cultivated a meaningful connection to what
I'd only ever regarded as vaporous arrangements of nothingness. I
wanted in. Also, I was impressed that these enthusiasts seemed to
be rattling the self-serious strictures of the scientific establishment.
And so it was disappointing to realize that nothing was really happening yet and that no one seemed particularly rattled.

Eventually Pretor-Pinney even sounded slightly exhausted by
asperitas. "It's the zombie news story that will never die!" he said.
He was by then closing in on his 10th year as head of the Cloud
Appreciation Society, and as he'd done after 10 years with *The Idler*
magazine, he was questioning his commitment to it. Somehow
being a cloud impresario had swallowed an enormous amount
of time. He was lecturing about clouds around the world, sharing stages at corporate conferences and ideas festivals with Snoop
Dogg and Bill Clinton and appearing monthly on the Weather
Channel. Then there was the Cloud Appreciation Society's online
store, a curated collection of society-branded merchandise and
cloud-themed home goods, which turned out to be surprisingly
demanding, particularly in the frenzied weeks before Christmas.
The Cloud Appreciation Society was basically just Pretor-Pinney
and his wife, Liz, plus a friend who oversaw the shop part-time
and a retired steelworker he brought on to moderate the photo
gallery. It was all arduous, which Pretor-Pinney seemed to find a
little embarrassing. "My argument about why cloud-spotting is a
worthwhile activity is that it's an aimless activity," he said. "And I've
turned it into something that is very purposeful, that is work."

At the same time he realized that he'd conjured a genuine community of amateur cloud-lovers from all over the world but regret-
ted never doing anything to truly nourish it; it felt so "fluffy," he
said, "with no center to it, like a cloud." Soon that spectral society
—that cloud of people on the Internet—would be celebrating its
10th anniversary. "I'm thinking that it might be a nice reason to
get everyone together," he said.

One morning last September, Pretor-Pinney was fidgeting and
fretting in the auditorium of the Royal Geographical Society
building, at the edge of Kensington Gardens in London. "Escape to the Clouds," a one-day conference to celebrate the Cloud

Appreciation Society's 10th anniversary, would be underway in 90 minutes, and Pretor-Pinney was impatiently supervising the small team of balloon-installation artists he had commissioned to rig inflatable cloud formations around the stage. This was the first big event that he organized for the Cloud Appreciation Society. The evening before the conference, he was expecting 315 attendees. But there was a late surge of ticket-buying, and now he was panicking about running out of artisanal Cloud-Nine Marshmallows for the gift bags. Outside, Pretor-Pinney kept pointing out, the London sky was impeccably blue. Not a single cloud. It was terrible.

Bounding onstage to kick off the conference, Pretor-Pinney seemed overwhelmed but cheerful. He reminded the muddle of cloud appreciators from all over the world, now crammed into the theater, that "to tune into the clouds is to slow down. It's a moment of meteorological meditation." And he celebrated the transcendence of cloud-spotting: how it connects us to the weather, the atmosphere, to one another. "We are part of the air," he told everyone. "We don't live beneath the sky. We live within the sky."

Who were they all? Why were they there? They were a collection of ordinary people with an interest in clouds. Behind all those usernames on the Cloud Society website were schoolteachers, skydivers, meteorologists, retired astronomy teachers, office workers, and artists. Many people had come alone, but conversations sparked easily. ("I've just seen the best cloud dress I've seen in my life," a woman said on the stairway. A second woman turned and said, "Well, yours is quite lovely too.") The atmosphere was comfortable and convivial and amplified by a kind of feedback loop of escalating relief, whereby people who arrived at a cloud conference not knowing what to expect recognized how normal and friendly everyone was and enjoyed themselves even more.

The program Pretor-Pinney had pulled together was a little highbrow but fun. A British author recounted the misadventures of the first meteorologist to make a high-altitude balloon ascent. An energetic literary historian surveyed "English Literary Views of the Sky." Pretor-Pinney and a professor of physics tried to demonstrate a complicated atmospheric freezing process in a plastic bottle, but failed. And between the talks a musician named Lisa Knapp performed folk songs about wind and weather. She had saved the obvious crowd-pleaser for her final turn onstage: the melancholy Joni Mitchell classic "Both Sides, Now."

There would be one more talk after Knapp finished, but it didn't matter. This—the Joni Mitchell moment—was the conference's transformative conclusion. Knapp had an extraordinary voice, Bjork-like but gentler, and performed the song alone, accompanying herself with only a delicate, monotonal Indian classical instrument resting in her lap, a kind of bellows, called a shruti box. It let out a mournful, otherworldly drone. After hours of lectures and uncertain socializing with strangers, something about this spare arrangement and the sorrowful lyrics felt so vulnerable that by the time Knapp finished the first lines—"Rows and floes of angel hair . . . I've looked at clouds that way"—she was singing into an exquisite silence.

The performance moved me. But it was more than that, and weirder. Maybe somewhere in this story about clouds and cloud lovers I'd found a compelling argument for staying open to varieties of beauty that we can't quite categorize and, by extension, for respecting the human capacity to feel, as much as our ability to scrutinize the sources of those feelings. Whatever the case, as Knapp sang, I started to feel an inexplicable rush of empathy for the people I met that day, the people sitting around me—all these others, living within the same sky. And I let my mind wander, wondering about their lives. What I felt, really, was awe: the awe that comes when you fully internalize that every stranger's interior life is just as complicated as yours. It seemed very unlikely that a meeting of an online cloud society in a dark, windowless room could produce such a moment of genuine emotion, but there I was, in the middle of it. Just thinking about clouds, I guess, had turned a little transcendent, at least for me.

Then I heard the sniffle. It was very loud. With the room so transfixed, it easily cut through Knapp's voice from a few rows behind me, and when I turned to look, I saw Pretor-Pinney's wife fully in tears. Then the woman right next to me, she was crying too. And I heard others inhaling loudly, oddly, and got the impression there were more. Immediately afterward, out in the hall, the first person I walked past was bashfully apologizing to two others. It was so strange, she kept saying. She just didn't know why she'd been crying.

A couple of days later I tried to describe it in an email to a friend: "Many people spontaneously cried, just releasing their tears like rain, and I realized that we are all human beings—that's the truth . . . in all our different forms and sizes, we are expressions of the same basic currents, just like the clouds." And when I read the email back, I was mortified by how fluffy and stoned it sounded, but still—even now—I can't pretend it's not true.

MICHAEL REGNIER

The Man Who Gave Himself Away

FROM *Mosaic*

LAURA MET GEORGE in the pages of *Reader's Digest*. In just a couple of column inches, she read an abridged version of his biography and was instantly intrigued. In the 1960s, apparently, egotistical scientist George Price discovered an equation that explained the evolution of altruism, then overnight turned into an extreme altruist, giving away everything up to and including his life.

A theater director, Laura Farnworth recognized the dramatic potential of the story. It was a tragedy of Greek proportions — the revelation of his own equation forcing Price to look back on his selfish life and mend his ways, even though choosing to live selflessly would lead inexorably to his death. But as she delved into his life and science over the next five years, Farnworth discovered a lot more than a simple morality tale.

Born in New York in 1922, George Price realized pretty early on that he was destined for greatness. In a class full of smart kids he was one of the smartest, especially with numbers. He was in the chess club, obviously, and his mathematical brain was naturally drawn to science. Determining that there was no rational argument for God's existence, he became a militant atheist too.

His PhD came from the University of Chicago for work he did on the Manhattan Project — having graduated in chemistry, he'd been recruited to find better ways to detect traces of toxic uranium in people's bodies. Although it had been a top-secret project, young Price must have felt he was already part of world events. Obsessed with applying his brilliance to big problems, however, he struggled to find a job that satisfied him. Instead he pursued his big ideas outside work, and not only scientific ones: he wasn't afraid of wading into public arguments with famous economists, and even sent his plans for world peace to the U.S. Senate. He didn't understand why other people didn't take up his ideas: the solutions seemed so obvious to him.

Domestic problems were a different matter. He'd met his wife, Julia, on the Manhattan Project, but as well as being a scientist she was a devout Roman Catholic. The marriage was hard-pressed to survive Price's scathing views on religion, and after eight years and two daughters—Annamarie and Kathleen—they divorced. Fed up with his job, his life, and the distinct lack of recognition in America, Price cut his ties in 1967 and crossed the Atlantic to London, intent on making a great scientific discovery there. He felt he had just a few more years to make his mark, but as it turned out, he needed only one.

Price had set himself the "problem" of explaining why humans lived in families—particularly what fatherhood was for, scientifically speaking. This in turn led him to the question of how altruism had evolved, and it was while studying new theories around this topic that he derived what is now called the Price equation, almost by accident.

This is what it looked like:

$$w\Delta z = \text{cov}(w_i, z_i)$$

It captured the essence of evolution by natural selection in one simple formula. It describes how in a population of reproducing individuals, be they people, plants, or self-replicating robots, any trait (z) that increases fitness (w) will increase in the population with each new generation; if a trait decreases fitness, it will decrease. It's a type of statistical relationship called covariance, and it was so elegant that Price couldn't quite believe no one had stumbled across it before.

So in September 1968, this obscure middle-aged American scientist walked in off the street to the Galton Laboratory, the home of human genetics at University College London. No one there knew who he was—he had no credentials, held no academic position, and had no appointment. All he had was an equation. When he confidently proclaimed in his condescending, high-pitched voice that his equation could explain the evolution of altruism, they probably thought he was a crank. Nevertheless, when he walked out 90 minutes later, Price had a job and the keys to his own office.

He continued to hone his equation there, but at the same time began giving away his possessions. He would seek out the homeless in Soho Square or at the nearest railway stations, Euston and King's Cross, and give them anything they asked for, from the money out of his pay packet right down to the clothes off his back. If they needed a place to sleep, he would invite them back to his flat indefinitely. Eventually he had given away so much that he became as destitute as the men he was helping. When the lease ran out on his flat, he took to squatting, moving often, somehow continuing to do research as well.

By the end of 1974, Price had given up everything. Sometime before dawn on January 6, 1975, in a squat not far from Euston, he killed himself.

Told like that, it seems obvious that everything was connected—he studied the concept of family because of the way he'd left his wife and daughters; his subsequent altruism was related to the equation he discovered; his suicide was a result of his extreme altruism. But as Farnworth discovered, nothing in Price's story is that simple.

To understand the sequence of events in his life, she set about drawing up a timeline based on his letters (archived in the British Library), the 2010 biography of Price that had prompted that short piece in *Reader's Digest,* and other sources.

Knowing more of the details changes the story. For example, despite the implication that he deserted his daughters, they never felt he had abandoned them. Kathleen's attitude is that it was normal in the 1950s for children to stay with their mother after a breakup, plus their father had remained a part of their lives, taking them to museums, concerts, and the theater. Yes, they saw less of him when he had to move away for a new job, but in her late teens Kathleen spent some time in New York, not far from where Price was then living, and she has fond memories of long walks through the city together, his love of poetry and Shakespeare, and his insatiable intellectual curiosity.

In 1966, more than a decade after the divorce, Price needed an operation to remove a tumor that had been lurking in his thyroid for a few years. Fatefully, he asked an old friend to do the surgery, and while removing the entire thyroid gland cured the cancer, it had serious consequences for Price's health. A nerve in his right shoulder was damaged in the operation, leaving him extremely bitter about his (former) friend's "butchery" and without feeling in his arm and on one side of his face. In addition he had to take thyroxine pills to replace the hormones his thyroid used to make. On occasion Price would stop taking his pills and experience profound episodes of depression as a result.

On a more positive note, Price's medical insurance paid out handsomely, and it was this money that funded his move to London. Far from abandoning Annamarie and Kathleen, by then 19 and 18 years old, respectively, he stayed in touch, writing often. But conscious of his own mortality, he felt time was running out and that by moving away he would be able to focus on one brilliant, final piece of research.

It's inconceivable that his choice of family as a topic was not bound up with his relationship with his children, but the evolution of social behavior—and of altruism in particular—was also one of the biggest scientific questions of the age. It was threatening to undermine Darwin's whole theory of evolution by natural selection, which made it more than worthy of Price's obsessive attention.

Altruism has always been a bit of a problem. Every altruist has their own motives, of course—some are emotional, responding to fellow humans in desperate straits, while others are more rational, thinking about the kind of society they'd like to live in and acting accordingly. Does that imply a level of self-interest? Even if it did, it shouldn't undo the goodness of altruism, and yet people can be deeply suspicious of those who apparently willingly put others' interests before their own. Selfless acts often attract accusations of hidden selfishness, suggesting they're not really altruistic at all.

This wasn't the problem for Darwinism. After all, humans have culture and religion and moral codes to live by—maybe our altruism was more to do with that than biology. Unfortunately, altruism was not only a human trait—it was everywhere. There were birds that nurtured other pairs' fledglings, vampire bats that regurgitated blood for those who'd failed to feed in the night, monkeys that put themselves in danger by raising the alarm when a predator approached the rest of their troop.

It was altruistic ants that posed a particular problem for Charles Darwin. Natural selection is often described as "survival of the fittest," where fitness means how successful an individual is at re-producing. If one individual has a trait that gives them a fitness advantage, they will tend to have more offspring than the others; because the advantage is likely to be passed on to their offspring, that trait will then spread through the population. A fundamental part of this idea is that individuals are competing for the resources they need to reproduce, and fitness includes anything that helps an individual reproduce more than the competition.

But as Darwin observed, ants and other social insects are not in competition. They are cooperative, to the extent that worker ants are sterile and so have literally zero fitness. They ought to be extinct, yet there they are in every generation, sacrificing their own reproductive ambitions to serve the fertile queen and her drones. Darwin suggested that competition between groups of ants—queen, drones, and workers together—might be driving natural selection in this case. What was good for a nest competing against other nests would then outweigh what was good for any individual ant.

Group selection, as this idea was known, was not a very good solution, though. It didn't explain how the cooperative behavior evolved in the first place. The first altruistic ant would have been at such a huge disadvantage compared to the rest of its group that it would never have got the chance to breed more altruistic ants. The same was true of humans—natural selection was intrinsically stacked against any altruistic individual surviving long enough to pass on their altruism.

This left a rather embarrassing paradox: the evolution of altruism was impossible, yet clearly altruism had evolved. If the biologists couldn't resolve this, would they have to throw out the whole idea of natural selection?

Luckily, a young man called Bill Hamilton spared biology's blushes with a slightly different solution in 1964. He proposed that altruism could have evolved within family groups—yes, an individual altruist would seem to be at a disadvantage, but that was not the whole picture, because other individuals who shared the same genes associated with altruism would all influence each other's "inclusive fitness."

Discussions of human altruism are often framed in terms of someone drowning in a pond. Do you put your own life at risk to try and save them? If you do, that's altruism. Hamilton's idea, which became known as kin selection, acknowledged that compared to a selfish person who never got their feet wet, someone who went around jumping into ponds to save drowning people would be at a greater risk of dying before they managed to reproduce and pass their altruistic genes on to their children. However, if they happened to save a relative who shared the same genes, our altruist would have indirectly helped to get those genes passed on to the next generation after all. If the total benefit derived from having altruistic genes in the family, so to speak, was greater than the cost, then the evolution of altruism was no longer paradoxical.

When George Price stumbled across Hamilton's work in the Senate House Library in 1968, he was shocked. He was forced to confront the relationship between morality and family, the biological imperative he should have felt to sacrifice his selfish ambitions in favor of supporting his kin. He immediately set to work to challenge, even disprove, Hamilton's theory. But he could only confirm it. Along the way he derived his equation of natural selection, which helped to prove that altruism was not selfless and moral but rather selfish and genetic.

Laura Farnworth wanted to be a dancer when she grew up. When scoliosis put an end to dreams of ballet school she turned to theater instead, but like Price, her ambitions were stymied by poor health. In Farnworth's case, ulcerative colitis and a subsequent MRSA infection put her out of action for four years just as she had started to make her mark as a theater director. In 2011, when she had at last started working again, the idea of making a play about Price was also, therefore, about putting herself back center-stage.

Ambition was tempered with respect, however—rather than play up to the obvious version of his story, Farnworth wanted to "do right by George," which meant digging deeper into the true meaning of his actions and his research. But, she admits, understanding the Price equation was a constant struggle: "It's like juggling with three balls. I can juggle with two, but throw the third one at me and I drop it. I get two parts of the equation, but when I try and understand the third bit, I lose it."

Price added the third bit while employed at the Galton Lab. Here's what the next version looked like:

$$w\Delta z = \text{cov}(w_i, z_i) + \text{E}(w_i \Delta z_i)$$

The new bit on the right-hand side accounts for any effects the trait in question might have on its own transmission—if it has properties that make it more likely to be passed on than other traits. Having this extra term opened up the process to allow for more than the simple story of "survival of the fittest"—this was where Hamilton's ideas of inclusive fitness and kin selection could start to influence the course of evolution. It even allowed group selection more broadly; indeed, Price thought it meant natural selection could occur at many levels simultaneously. He wrote to Hamilton at once.

In his memoirs, Hamilton isn't sure when he and Price became friends; reading their letters 40 years later, Farnworth was able to see their friendship develop more clearly. Price had written first, within days of reading the papers on kin selection. Hamilton had replied politely enough, not suspecting what was to come, then had gone to Brazil on an extended field trip, so there was a gap in their correspondence of about a year. During this time, prompted by an offhand remark in Hamilton's letter, Price had toyed with applying game theory to biology (this work, taken up and developed by other scientists, actually helped move evolutionary biology forward much more than his equation). But he'd also set about working to improve the math of the kin selection theory, making it "more transparent." By the time Hamilton was contactable again, Price had derived the first version of his equation and got his job at University College London. But it was the extended version of the equation that he really wanted his new friend to see.

Today some scientists will tell you that the Price equation is empty. It is like a footballer who, when asked how their team will win the next match, says they will score more goals than the other team. By trying to explain the game at its most fundamental level, say the critics, the equation explains and predicts nothing about why certain traits should increase or decrease fitness.

In her own quest to crack its meaning, Farnworth went to ask three evolutionary biologists who think there is more to it than that. One told her it was "quite simple, really." Another jumped up and started scribbling diagrams and equations on the whiteboard in his office. The third said, "None of us understands it really; it resonates in context." For its supporters, the Price equation is the closest thing biology has to $E=mc^2$. It is a fundamental expression of natural selection that can be used to clarify concepts, separate different components of selection, and compare more specific mathematical models of evolution.

As for Hamilton, he was delighted with it from the moment he saw it. The Price equation was not, as Price had hinted to him, a new derivation or correction of his ideas. Instead it was "a strange new formalism that was applicable to every kind of natural selection." Its strangeness came precisely because Price was not a biologist—instead of starting from the work of their scientific forebears, he had worked everything out for himself from first principles.

"In doing so," wrote Hamilton, "he had found himself on a new road and amid startling landscapes."

In June 1970, just a few months after Hamilton had written to say how "enchanted" he was with the equation, Price had a profound religious experience (he refused to tell his friend the details, sensing that Hamilton would be as unbelieving of such things as Price himself would have been until that point). Depressed, apparently by his role in confirming that altruism had selfish origins —though it is just as likely that he had stopped taking his thyroxine pills again—he had become obsessed with coincidences in his life, not least the sheer improbability that he, who hadn't known "a covariance from a coconut," should have discovered that equation. All at once, despite a lifetime of hardline atheism, he became convinced that a higher power had been at work.

The nearest church was All Souls, just above Regent Street in central London. He walked in and started praying. By the time he walked out again, he had given himself to Jesus.

At first he brought the full weight of his intellect to bear on the Bible—he concluded that Easter week had taken 12 days, not eight, and was determined to persuade others of this truth, writing arguments as rigorous and detailed as his scientific research. Typical Price: start from first principles, take nobody else's word for it, test it obsessively, and try to find your own way to the truth.

Then, at the end of 1972, he had a second conversion. He had already decided to trust in Jesus completely. He'd stopped taking his thyroxine pills and by now his insurance money was starting to run out—but if Jesus wanted to save him, Jesus would find a way. Around Christmas he collapsed, close to death. A neighbor found him and he was rushed to hospital, where the doctors saved his life. For Price this was a sign that Jesus did want him to live, but also to change his ways and stop worrying about the length of Holy Week. He told Hamilton he had "sort of 'encountered' Jesus." He had had a vision, in other words, and heard Jesus whisper, "Give to everyone who asks of you."

Not everyone approved. A friend advised him against trying to "out-God God," while even the vicar at All Souls said giving money to down-and-outs was "seldom more than an easy way out for ourselves." But Price carried on giving away his worldly possessions whatever the consequences. He would even give away the big aluminum cross he wore round his neck if someone asked for it. Giving had become a compulsion, an addiction.

Farnworth turned to Isabel Valli at the Institute of Psychiatry, King's College London, to analyze Price's letters for clues to his mental state during this period. The meticulous detail in his letters suggested they were reliable accounts of events, but to Valli they were like a psychiatric interview. She got an insight into the way his thought processes were changing while he was in London. Linearity gradually gave way to circular thinking; he would go off on tangents, spiraling ever further away from what he was trying to say. There was a logic to it, but one that became harder for anyone else to follow.

Price was probably experiencing psychotic delusions, paranoia, and hallucinations beyond his visions of Jesus, not to mention depression exacerbated by thyroid hormone deficiency. According to Valli, it is human nature when trying to make sense of delusions to construct explanations based on things already significant in our lives; for Price, those things were religion and altruism (and also marriage—he proposed to several women around this time, including suggesting to Julia that they get remarried; like the others, she declined).

It's not that his altruism was a symptom of mental illness, nor that his equation turned him into an altruist; it was just another part of his increasingly disordered life that he was trying to incorporate into a consistent worldview.

For him, the most rational explanation available was that he had been chosen by God to discover the Price equation and to become an extreme altruist. He was happy to tell people about it, too—if he is remembered at all, one of the first things people tell you about him is that he ran through the corridors of University College London shouting that he had "a hotline to Jesus." In some ways his life had become extremely complicated, but it was also much more simple to be willing to give up anything and everything and put all his faith in Jesus.

Of course, that isn't the end of the story. As Farnworth sees it, Price had a third conversion shortly before he died. He finally stopped helping others. He didn't have much more to give by this point, it's true, but he began to pay more attention to his own well-being. He had realized that he needed to help himself first if he was going to be any use to anyone else. Rebuilding his life from the bottom up was a daunting task, however. Despite getting a job as a cleaner at a bank, he knew he was struggling. He made an appointment to see a psychiatrist. But then, just days before his appointment, he killed himself.

George Price was buried in an unmarked grave in St. Pancras Cemetery, a few miles north of central London. Bill Hamilton was at the funeral service, along with some of the homeless men Price had helped. Afterward Hamilton went to the squat where Price had been staying to collect any scientific papers he had been working on.

"Although the house was awaiting demolition the electricity was still on: it mightn't have been too freezing for George when he was there all alone over Christmas," Hamilton wrote in his memoirs. "As I tidied what was worth taking into the suitcase, his dried blood crackled on the linoleum under my shoes: a basically tidy man, he had chosen to die on the open floor, not on his bed.

"That is how his life became dreamlike for me and also how his colourful thread in my science and my life ran out."

On March 29, 2016, Farnworth's play *Calculating Kindness* opened for a sold-out three-week run at the Camden People's Theatre. It's a small community venue not far from where Price lived, worked, and died (though in his day it was the Lord Palmerston pub).

Given that Price didn't selfishly desert his family, his equation wasn't strictly about altruism, and his altruism didn't directly cause his death, Farnworth had to make some choices about the story she was going to tell. She could stick to the morality tale with its inherently simplistic drama, or she could trust in the compelling tragedy of his life in all its complexity. In the end she decided the latter would be the right thing to do by George. She would present his shifting worldview and let the audience draw their own conclusions. As such, the play doesn't offer any neat answers; it doesn't tell anyone how to live, or how selfish or altruistic we should be. Like the Price equation, it describes what happened, not what will or ought to be.

Among the audiences that came to see the show were Annamarie and Kathleen Price. Farnworth had invited them over from America but was still incredibly nervous about what their reaction would be. They were delighted. Kathleen said she saw the essence of her father onstage.

While in London, the sisters took the opportunity to arrange for a headstone to be placed at their father's grave. Below his name, it reads: "Father. Altruist. Friend." And there at the bottom, the Price equation, engraved in stone.

SONIA SMITH

Unfriendly Climate

FROM *Texas Monthly*

ONE CLEAR DAY last spring, Katharine Hayhoe walked into the limestone chambers of the Austin City Council to brief the members during a special meeting on how prepared the city was to deal with disasters and extreme weather. A respected atmospheric scientist at Texas Tech University, the 43-year-old had been invited to discuss climate change, and she breezed through her PowerPoint slides, delivering stark news in an upbeat manner: unless carbon emissions were swiftly curbed, in the coming decades Texas would see stronger heat waves, harsher summers, and torrential rainfall separated by longer periods of drought.

"Why do we care about all of this stuff?" Hayhoe asked. "Because it has huge financial impacts." The number of billion-dollar weather disasters in the United States had ballooned from one or two per year in the '80s to eight to 12 today, Hayhoe explained as she pulled up a slide with a map of the country. "Texas is in the crosshairs of those events, because we get it all, don't we? We get the floods and the droughts, the hailstorms and the ice storms, and even the snow and the extreme heat. And we get the tornadoes, the hurricanes, and the sea-level rise. There isn't much that we don't get."

Soon afterward Don Zimmerman, a conservative councilman who before being elected regularly sued the city over tax increases, declared from his seat on the dais that climate change was a "nebulous" and "foolish" field of study. Zimmerman, wearing a banker's collar and projecting an officious air into the room, continued, "We have maybe 30 years of satellite data, and the world is maybe millions of years old. I have a really visceral reaction against the climate-change argument, for the simple reason that when you look back in time, there have been dramatic climate changes before humanity ever existed.

"The worst thing that can be done to humanity is put government bureaucrats in charge of carbon dioxide emissions," he said as Hayhoe listened politely. "You don't have to be as smart as a fifth grader to know that what causes the climate is the sun. I have people tell me, 'Carbon dioxide warms the earth.' No, it doesn't. The sun warms the earth, and there is more energy in our sun than humanity can comprehend." Zimmerman then insisted that the sun didn't need "a permit from the EPA" to emit solar flares.

An uncomfortable silence settled over the chamber for a moment before Hayhoe joked, "I think if the EPA could be in charge of the sun, that could create bigger problems than we have today." She then proceeded to gut Zimmerman's arguments. "A thermometer is not Democrat or Republican, and when we look around this world, it's not about trusting what our 30-year-old satellites say. It's about looking at 26,500 indicators of a warming planet, many of them we can see in our own backyards," she said. The climate was not changing because of orbital cycles, which bring about ice ages, Hayhoe maintained. "The earth's temperature peaked 8,000 years ago and was in a long, slow slide into the next ice age until the industrial revolution," she said. Instead of being in this cooling period, the planet had seen its average temperature steadily rise. The sun was also not the culprit: "If the climate were changing because of the sun, we'd be getting cooler, because energy from the sun has been going down over the last 40 years," she said.

But Zimmerman, it seemed, had no use for facts, and after the meeting he continued to harangue Hayhoe. The encounter, however, came as no surprise. In fact, it was depressingly familiar to Hayhoe, who has auburn hair, hazel eyes, and a calm, affable nature that is reminiscent of an excellent physician's bedside manner. And she often likens herself to a doctor, but her patient is the planet. After taking its temperature, she feels compelled to report her diagnosis: because of manmade carbon emissions, the earth is running a fever. She knows that this message doesn't always find a receptive audience. Over the past 15 years, climate change has emerged as one of the most polarizing issues in the country, ahead of guns, the death penalty, and abortion. And there is no group

that is more unconvinced of climate change's reality than evangelical Christians, who primarily identify as conservative Republicans.
As Brian Webb, the founder of the faith-based Climate Caretakers,
recently told Religion News Service, "The United States is the only
industrialized country in the world where denial of climate change
has become inextricably linked to a dominant political party."

All of which puts Hayhoe in a unique position. A coauthor
of the last two National Climate Assessments and a reviewer on
the Nobel Prize–winning Intergovernmental Panel on Climate
Change, Hayhoe—the daughter of missionaries and the wife of
a pastor—is herself an evangelical Christian. In her talks she uses
the Bible to explain to Christians why they should care about climate change and how it affects other people, from a poor family
on the island nation of Kiribati who will be displaced by rising sea
levels to an elderly couple in Beaumont who can't afford to pay
for air conditioning in Texas's increasingly sweltering summers.
As she puts it, "The poor, the disenfranchised, those already living
on the edge, and those who contributed least to this problem are
also those at greatest risk to be harmed by it. That's not a scientific
issue; that's a moral issue."

Hayhoe maintains a dizzying schedule. In the past year she has
attended the historic United Nations climate summit in Paris, traveled to the edge of Hudson Bay, in Canada, to witness the annual
polar bear migration, curated a special *Good Housekeeping* issue on
climate change, and appeared onstage in New York with Gloria
Steinem at a talk at the Rubin Museum of Art. That's in addition
to teaching her graduate-level seminars, serving as a codirector
of Texas Tech's Climate Science Center, and publishing 17 scientific papers. (Travel is essential for Hayhoe's job, but to do her
part—and perhaps head off criticism about her carbon footprint
—Hayhoe buys carbon offsets to reduce the impact of her trips.)
One warm afternoon in October, on a day spent in Lubbock between visits to Colorado and Houston, Hayhoe spoke at a Phi Beta
Kappa ice cream social inside Texas Tech's Hall of Nations, a room
draped with the flags of 190 countries and featuring a glossy ter-
razzo map of the world on the floor. The crowd, mostly professors
from across the university's departments and a smattering of students, dug into Styrofoam bowls of vanilla and cookies and cream

as Hayhoe, who was wearing a red top and flowing linen pants, began her speech.

"I'm a professor here at Tech, and what I'm going to talk about today is not my research. I'm going to talk about the experience that I have talking about my research. Now, most of you are not going to have the same experience I do. If you study literature, you don't have to spend a lot of time convincing people that books are real. If you study engineering, most people will agree that engineering is real and it's an important part of our society. But I study something that about half of the country and much more than half of Texas thinks is a complete hoax," she said. "Many people view having climate science at Texas Tech as similar to having a Department of Astrology. But we don't use crystal balls, we use supercomputers; we rely on physics, not brain waves."

The study of climate science dates to 1824, when French physicist Joseph Fourier discovered what would become known as the greenhouse effect, in which gases trapped in the atmosphere absorb heat and raise the temperature of the planet. It took 35 more years for John Tyndall, an Irish chemist, to pinpoint carbon dioxide as one of the heat-trapping gases in the earth's atmosphere. And in 1896 a Swedish chemist named Svante Arrhenius declared that burning coal contributed to the greenhouse effect, after spending almost two years calculating (by hand!) how increasing carbon dioxide concentrations raised the earth's temperature. So the basic science, as Hayhoe often points out, has been settled since before the start of the 20th century. Today there is robust scientific consensus that global warming is "real, caused by humans, and dangerous"; a study found that 97 percent of climate scientists agree with those conclusions. The Department of Defense calls climate change a "threat multiplier," because it exacerbates existing problems. And the year 2015 was the warmest on record, breaking the previous mark, which was set in 2014.

So why is climate science greeted with so much skepticism? Part of the reason can be attributed to the way the topic is often handled in the media. On cable news, two people from opposite sides of the debate are typically paired to argue about the subject, but that can lead to a false equivalency between scientists on the one hand and paid spokesmen on the other. As historians Naomi Oreskes and Erik M. Conway chronicled in the book *Merchants of*

Doubt, some of the most prominent climate-change skeptics are the same politically conservative scientists who were previously funded by Big Tobacco to spread falsehoods about cigarettes. Their employer this time around? The fossil-fuel industry.

And part of the reason is the suspicions that conservatives have of government intervention. Hayhoe has found that some people don't reject the reality of climate change because they disagree with the science but because they fear that the solutions will upend their lives. This seems to be the case for U.S. senator James Inhofe, a Republican from Oklahoma, who once told journalist Rachel Maddow, "I thought it must be true until I found out what it would cost." That day at Tech, Hayhoe recounted an anecdote about an experience she'd had speaking to a group of water managers for the Brazos River a few months back. At the end of that talk, an older man stood up and said, "Everything you said makes sense, but I don't want the government telling me where to set my thermostat."

Some critics feel so threatened that they resort to ad hominem attacks on climate scientists. Hayhoe receives a steady stream of hate mail, which she files away in a special folder. When I asked her when this started, she replied, "The first time I was ever quoted in a newspaper article." The ugliness reached its height in 2012, during the presidential race. At the time Hayhoe was writing a chapter on global warming for a book Republican hopeful Newt Gingrich was coauthoring about the environment. Rush Limbaugh mentioned it on his radio program, dismissively referring to Hayhoe as a "climate babe." A few days later an Iowa voter buttonholed Gingrich on camera to ask him about it, and Gingrich swiftly replied, "That's not going to be in the book. We didn't know that they were doing that—we told them to kill it." Hayhoe took to Twitter to respond: "What an ungracious way to find out, eh? Nice to hear that Gingrich is tossing my #climate chapter in the trash. 100+ unpaid hrs I cd've spent playing w my baby."

Most of the time she laughs these incidents off. "I got one today that was exceptional," she told me in late September, as we sat inside the Climate Science Center. "Most of the stuff is rambling, but this one was not. Someone wrote on Facebook, 'She is a lying lunatic, and probably a witch.' That was very concise," she said with a grin. But sometimes the comments veer into violent territory. Hayhoe recalls one email that prompted her to call authorities.

"You are a mass murderer and will be convicted at the Reality TV Grand Jury in Nuremberg, Pennsylvania," the email began. "After the Grand Jury indicts you, I would like to see you convicted and beheaded by guillotine in the public square, to show women that if they are going to take a man's job, they have to take the heat for mass murder." But most of the time Hayhoe doesn't let such vitriol drive her to despair, though dealing with it can be exhausting. "What frustrates me the most, and what I find difficult not to take personally, is how much of the hate mail comes from so-called Christians."

That bile is something Hayhoe never anticipated when she was applying to graduate school 22 years ago. A native of Toronto, she had double-majored in physics and astronomy at the University of Toronto and spent every clear night one summer gazing through the telescopes on top of the physics building. She found that the astronomer's life appealed to her and planned to study that in graduate school. Then she took a climatology class her junior year. "Until I took that course, I did not realize that climate change is affecting everything, from poverty to biodiversity to health, and so you can't fix any one of them if you leave climate change out of the picture," she told me. She also realized that her background in physics had perfectly positioned her to study climate modeling.

If she was going to leave astronomy behind, Hayhoe wanted to do policy-relevant climate science. When she was considering graduate programs, she was thrilled to learn that Don Wuebbles, who had been instrumental in addressing the chlorofluorocarbon problem in the '80s, was the new head of the Department of Atmospheric Sciences at the University of Illinois, Urbana-Champaign. He would serve as her adviser for both her master's degree and her doctorate. Under Wuebbles's guidance, Hayhoe eventually began focusing on statistical downscaling, which was still a relatively new field when she started graduate school, in 1995. "There was very little of this being done at the time," Wuebbles recalled recently, "and the methods were not capturing the full extent of the science, so she set about to develop a new technique and very successfully did so. She's brilliant."

Statistical downscaling involves combining historical weather observations with global-climate models to better predict what the future could look like in a particular place. "The local environment, whether it's hilly or flat, with crops or forest, urban or rural, modifies the weather patterns we get," she said. "So, for example, if we had identical high-pressure systems over Lubbock and Houston, it would mean something different for the temperature, for the humidity, for the rainfall patterns." Hayhoe also tries to see if the global models reflect real-world conditions on the ground. "When we get an El Niño, we see a very wet winter from here in Lubbock all the way across to Florida. Do the models pick that up or not? We need to know," she explained.

Hayhoe runs simulations on a supercomputer, then she combs through the data to interpret the output. On a practical level, this means Hayhoe exists in a world of numbers, thousands upon thousands of lines of them. A single file dealing with one variable—say, temperature across the country over the next hundred years—can be almost five gigabytes in size. And she runs these simulations for multiple variables and scenarios on multiple climate models. (Some 42 global-climate models exist today, run by labs around the world.) These reams of data are shapeless until she translates them by writing code. "What a lot of people don't realize is that the most important skill any climate scientist has is programming," she told me over pizza in Lubbock one afternoon last fall.

Hayhoe has used downscaling in her consulting work for the cities of Washington, D.C., Boulder, and Chicago as well as federal entities, including the Department of Defense and the U.S. Fish and Wildlife Service. She helps analyze problem areas, such as sewer overflow during heavy rain or warped train rails during heat waves, and tries to pinpoint how often those things will be a problem in the future, based on changing climate patterns. In 2004, Hayhoe was an author on a paper that examined California's future from different angles, from water supply to agriculture to tourism. She was heartened when, a few months later, that research prompted governor Arnold Schwarzenegger to sign an executive order limiting greenhouse-gas emissions. He was the first governor to do so. "When Schwarzenegger signed that bill, he had the authors from California standing in a semicircle behind him. The reason why I left astrophysics is to do policy-relevant research, and when I saw that picture, I thought to myself, 'I did it. This works.'"

Hayhoe's scientific credentials are impeccable, but what has made her an international star are her skills as a communicator.

John Abraham, an associate professor of thermal sciences at the University of St. Thomas, in Minnesota, has called her "one of the best climate communicators in the world." Abraham told me, "She is extraordinary at relaying very complex topics into language that other people can understand, without speaking down to them. The other thing she's good at is hearing questions. We all listen, but she has this innate ability to understand the perspective of the person making the inquiry," he said. "She has this knack for honestly presenting the science but doing it in a disarming way for people who are often antiscience."

One mild Friday in early October, I flew to Houston with Hayhoe and her eight-year-old son, who spent the short flight absorbed in the game Minecraft on his iPad while Hayhoe tapped away on her laptop. She was to give a keynote speech at Memorial Drive Presbyterian Church, a collection of limestone buildings nestled between pine trees in one of Houston's most affluent neighborhoods. The weekend symposium was called "Faithful Alternatives to Fossil Fuel Divestment." Hayhoe arrived with some tough talk for her audience. "There's no way to sugarcoat this, and I wish I could, because I know I'm in Houston, but the way that we get our energy does matter. If we continue to rely only on fossil fuels, we're going to end up on a very different pathway than if we gradually and sensibly transition to clean and renewable energy that we can grow here in Texas—and that many of our energy companies are already investing in very heavily."

The conference was organized in response to the Presbyterian Church (USA)'s proposal to divest church resources from fossil fuels, a move the Houston chapter had rejected as a symbolic one that unfairly vilified the people who work in the fossil-fuel industry. The group instead proposed that the national organization take steps to reduce its carbon footprint and advocate for a carbon tax. Hayhoe too is a proponent of putting a price on carbon and letting the markets sort it out. She thinks that a reasonable tax on gasoline would be around six cents a gallon. "Regulations just get more and more complicated, and you have to hire new people to deal with them," she explained. "It gets expensive and difficult to plot your strategy, but any business—from the ma-and-pa shop around the corner to the biggest multinational in the world—

knows what to do with a simple price change. Business is all about maximizing profit and minimizing costs. So in a sense, putting a price on carbon just frees up business to do what it does best."

But the most revealing part of her talk centered on why Christians should care about climate change. To lead into this subject, Hayhoe flipped to a slide with a quote from John Holdren, President Obama's science adviser: "We basically have three choices: mitigation, adaptation, or suffering. We're going to do some of each. The question is what the mix is going to be. The more mitigation we do, the less adaptation will be required and the less suffering there will be." *Suffering,* Hayhoe said, is not a word often deployed by scientists. "As scientists we don't know a lot about suffering, but as Christians we do. And we know that part of the reason we're here in this world is to help people who are suffering." And that suffering will not be meted out proportionally: if global warming continues unchecked, the poor—whether they're in Houston's Fifth Ward or in low-lying areas of Bangladesh—who have contributed least to carbon emissions will feel the most pain, from enduring more intense heat waves to paying the higher food prices that will accompany failed crops. Throughout the Bible, God charges Christians to serve others, Hayhoe said, from Genesis, where God makes man in his image so that he can be responsible for every living creature on earth, to 1 Peter 4:10: "'Each of you should use whatever gift you have received to serve others, as faithful stewards of God's grace in its various forms.'

"We've been given this commandment to love others as Christ loved us," Hayhoe said as a slide quoting John 13:34-35 flashed on the screen: "'Let me give you a new command: love one another. In the same way I loved you, you love one another. This is how everyone will recognize that you are my disciples—when they see the love you have for each other.'" She continued: "You can see, you just go through the Bible for verse after verse. They're not verses about climate change; they're not verses about the environment. They're verses about our attitudes and perspectives to other people on the planet. We are to be recognized for our love for other people." The members of the crowd nodded along in agreement as she spoke. The year 2015 was a good one to be proclaiming this message: in June, Pope Francis sent out his 192-page papal encyclical imploring the world's 1.2 billion Catholics to care about climate change, and in October the National Association of Evangelicals issued a similar call to action.

Hayhoe can speak honestly about suffering because of a lesson she learned when her parents became missionaries and moved the family to Colombia when she was nine. There she witnessed true poverty. Her father would travel to remote villages to speak at tiny churches, and she remembers hearing stories of landslides washing away homes after heavy rains. She now recognizes that these early memories of poverty and vulnerability have informed her work. Hayhoe was raised as a member of the Plymouth Brethren, a conservative, evangelical offshoot of the Anglican Church that emphasizes reading the Bible and interpreting it for oneself. This lent itself well to science, Hayhoe told me. "My dad was very much of the perspective that the Bible is God's first book and nature —creation—is God's second book."

Though Hayhoe has always been serious about her faith, connecting with groups of fellow Christians about climate change was not something she did before moving to Texas. In 2006 she and her husband, Andrew Farley, relocated from South Bend, Indiana, to Lubbock, one of the most conservative cities in the country, so that they could both take jobs at Texas Tech, he as a linguistics professor, she as a researcher. He also became the pastor at a small nondenominational church on the southwest side of town, now called Church Without Religion. People were surprised when they learned what the pastor's wife did, and Farley started getting lots of questions about it. And at Texas Tech the invitations for Hayhoe to speak about climate change started rolling in. The volume of these questions and the lack of resources to point people to spurred her and Farley to write a book together, *A Climate for Change: Global Warming Facts for Faith-Based Decisions.* The questions they tackle in the book were familiar territory for the couple, who had met through a Christian organization while in graduate school. A few months into their marriage, Hayhoe realized that Farley, who had grown up in a conservative household in Virginia, did not think climate change was real, and they began vigorously debating the topic. "It took about two years, but now we're on the same side," she said.

But beyond just speaking to Christian groups, Hayhoe prides herself on being able to talk to anyone with an open mind about the reality of climate change. She bemoans the fact that global warming has come to be viewed as a niche environmental issue. "To care about climate change, all you have to be, pretty much, is a human living on planet Earth. You can be exactly who you are with exactly the values you have, and I can show you how those values connect to climate change," Hayhoe told me.

Hayhoe's first step is always to "genuinely bond over a shared value," with an emphasis on that shared value's being genuine. "The key is not to pretend; we can all smell someone who is not genuine a mile away," she said. "If I'm talking to farmers or ranchers or water managers, I start off by talking about what we all care about, which is making sure we have water. And that, for many Texans, is almost as strong of a value as whatever it says in the Bible." Her next step is to connect that issue to climate change. So when talking about water, she describes how climate change is changing rainfall patterns. "We're getting these heavy downpours, and then we're getting longer dry periods in between, and our droughts are getting stronger because the warmer it is, the more water evaporates out of our lakes and rivers and our soil," she said. She tries to end her talks with solutions that inspire people, ranging from the personal (measuring your carbon footprint and installing energy-efficient light bulbs) to the large-scale (putting a tax on carbon). Hayhoe herself is most excited by the efforts of Elon Musk, the CEO of Tesla Motors and founder of SpaceX. "If I had to pick one person to save the world—and I don't think any one person will, but if I had to pick one—it would be him." She is excited about the battery packs that Tesla is developing, declaring energy storage the "single technology that will make the most difference."

Ultimately she does not care whether people agree with the science, so long as they take action. She compares this to a battle waged in the mid-1800s, before the germ theory of disease gained widespread acceptance, when a Hungarian physician urged other doctors to wash their hands and instruments before delivering babies. As doctors changed their habits, fewer and fewer women died from "childbed fever." "I don't care if they thought germs are imaginary, so long as they washed their hands," she said. The same is true for climate change, in Hayhoe's mind. If people start using more efficient light bulbs or driving more-fuel-friendly cars, it doesn't matter what they think about the science.

Hayhoe is coy about her own personal politics, and this air of mystery is useful to her. When I asked her about another Canadian-born Texan, climate-change skeptic and senator Ted Cruz, she demurred. She's a U.S. permanent resident but not a citizen, so she can't vote in the presidential election, and she seems to enjoy the level of remove this gives her from American politics. "It helps me not to pick sides, because people always ask if you're Democrat or Republican, and I'm neither. I can't be," she told me. "I appreciate the solutions that some Republicans are starting to advance, and I appreciate the fact that Democrats accept the science. But it's become so polarized that the good people on both sides are being marginalized." Whoever the next president is, Hayhoe hopes he or she will honor the commitments made at the climate summit in Paris last year and also put a price on carbon.

Hayhoe's religious background led NOVA's *Secret Life of Scientists and Engineers* to dub her a "climate-change evangelist" in 2011, and the label has stuck, though she is lukewarm on it. "An evangelist is someone who spreads good news, and I feel like I'm not really evangelizing. I feel more like a Cassandra, or an Old Testament prophet spreading bad news, saying, 'If thou dost not change from thy wicked ways and repent, thou shalt reap the harvest of thy deeds.'" But when Hayhoe talks, she doesn't sound so pessimistic. That's a strategic choice, as she realizes that doom and despair won't motivate others to act. For that, you need hope. "You have to offer people a vision of what the world could look like if we could wean ourselves off fossil fuels, if we could have a clean-energy economy," she said. "We would all want to live in that world."

Lyndon Baines Johnson was at his ranch outside Johnson City recuperating from gallbladder surgery on November 5, 1965, when his science advisers published a 317-page report warning about the dangers of air pollution. Tucked away in an appendix were 23 pages about atmospheric carbon dioxide. "Through his worldwide industrial civilization, Man is unwittingly conducting a vast geophysical experiment," the report states. "Within a few generations he is burning the fossil fuels that slowly accumulated in the earth over the past 500 million years." This additional carbon in the atmosphere would, over time, raise the earth's temperature, slowly melt the Antarctic ice cap, and lead to increased ocean acidity, the report proclaimed. "The climate changes that may be produced by the increased CO_2 content could be deleterious from the point of human beings," the report concluded.

Fifty years later Hayhoe gave the capstone presentation at a daylong symposium in Washington commemorating the first time a president was warned about the danger of climate change. "As several have already said today, we are conducting an experiment with our planet on a scale that has never before been attempted," she said, echoing the words of the report. The climate models that scientists now use churn out petabytes of data—which is something like, in Hayhoe's words, "twenty million four-drawer filing cabinets full of text"—that then need to be analyzed to see how these changes will manifest in particular locales. "What's the point of doing all of that modeling and all of that analysis if we don't understand how it's going to affect the system right here that we care about?"

Would LBJ even recognize the future Texas predicted by these models? In the past 50 years temperatures in Texas have risen half a degree per decade and are set to rise at least 3.5 degrees by midcentury if global emissions aren't slashed. "Our average summer could look like 2011 within my lifetime if we continue on our current pathway," Hayhoe told an audience in October, referencing that scorching summer when much of Texas saw more than one hundred 100-degree days. Austin could feel more like Scottsdale, Arizona. Rainfall patterns are shifting, so the state will face longer dry spells punctuated by more bouts of heavy rain. In West Texas, farming and ranching communities have thrived in the semiarid environment by pulling water from aquifers. But as the aquifers dry up, these communities are relying more on rainfall, just as that rainfall is becoming less likely and droughts are getting more intense, Hayhoe said. In LBJ's beloved Hill Country, this means increased risk of fire. Humans are the ones igniting the fires, but climate change is making them worse by providing the ideal dry conditions they need to spread. On the Gulf Coast, where a quarter of the state's 27 million people live, sea levels are already eight inches higher than they were a hundred years ago and are set to rise an additional one to four feet by the end of the century. And then there's the danger from stronger hurricanes fueled by record-breaking ocean temperatures.

Texas leaders, however, seem unwilling to tackle the problem or even admit that it exists. Governor Greg Abbott has long voiced skepticism about the science of climate change, telling the editorial board of the *San Antonio Express-News* during his gubernatorial campaign that the climate has always changed over time and further study was needed. "We must be good guardians of our earth, but we must base our decisions on peer-reviewed scientific inquiry, free from political demagogues using climate change as an excuse to remake the American economy," he told the newspaper. As attorney general, Abbott made a habit of suing the Obama administration, oftentimes over regulatory issues relating to climate change. His successor, Ken Paxton, is continuing that tradition, joining a lawsuit in October over the administration's Clean Power Plan, which calls on states to curb emissions by phasing out coal plants and shifting to natural gas and renewables. The plan would require Texas to decrease its coal power capacity by 4,000 megawatts, or 25 percent, and Paxton has likened this to the EPA's mounting a "war on coal and fossil fuels."

In such a milieu, efforts to incorporate climate change into planning at the state level have fallen flat, and bills that attempt to address it have gone nowhere in recent years in the legislature. "At the state level, in some circles, climate change is still a taboo subject," John Nielsen-Gammon, the state climatologist, told me. This leaves cities to do their own resilience planning. Meanwhile entities such as the Electric Reliability Council of Texas, the operator of the state's electric grid, are not taking climate change into account when developing their projections for load growth, which could lead to problems as the mercury creeps upward.

In Congress, Texans are some of the most vocal climate-change skeptics. Congressman Lamar Smith, a Republican from San Antonio, has used his chairmanship of the House Science, Space, and Technology Committee to tussle with federal agencies over their climate-change research, going so far as to subpoena the scientists who conducted a study with a conclusion he disagreed with and demand their emails. (Smith, it is worth noting, has received more than $600,000 in campaign donations from the fossil-fuel industry over his 29 years in Congress.) And then there's Cruz, who in December held a three-hour Senate hearing titled "Data or Dogma? Promoting Open Inquiry in the Debate Over the Magnitude of Human Impact on Earth's Climate," at which he claimed that there was a lack of scientific consensus on global warming.

Hayhoe is hopeful that as green energy gets cheaper, more people will begin using it. "Texas is unique, in that it is one of the states that have the most to lose economically from climate-change impacts, but Texas also has the most to gain by transitioning to a clean-energy economy," Hayhoe told me one day in her office on campus, a cluttered, windowless space. The room's sole decorative flair, a papier-mâché arctic fox that was a Christmas present from her young son, sat perched on a shelf.

If Texas were its own country, it would be the seventh most prolific emitter of carbon dioxide in the world. As it stands, Texas is the number one emitter in the U.S.; it released some 641 million metric tons of carbon dioxide into the atmosphere in 2013, almost double that of California.

But the state also has a seemingly boundless potential for green energy. Texas leads the nation in wind generation; turbines produced a full 10 percent of the state's power in 2015. By 2030 that number is forecast to jump to 37 percent. One night last September, supply of wind power was so plentiful and demand was so low that the spot price of electricity went negative for a few hours. Solar installation has lagged behind, but when it ramps up, there's enough capacity just in the hundred-square-mile area between Plainview and Amarillo to light the entire United States, as Hayhoe likes to point out. In Pecos County alone companies have plans to invest $1 billion in large-scale solar energy farms. "Texas understands energy. Energy is a Texas thing," Hayhoe told me. "We have the land we need to do this, as well as the technology and entrepreneurial spirit. I wish that the whole state could see that this is an opportunity for a better future."

EMILY TEMPLE-WOOD

It's Time These Ancient Women Scientists Get Their Due

FROM *Nautilus*

WOMEN ARE WOVEN deeply into the history of science, stretching back to ancient Egypt, over 4,000 years ago. But because their contributions often go unacknowledged, they fade into obscurity —and the threads of their influence today aren't as apparent as they ought to be.

As a Wikipedia editor, I have tried to make women's contributions more apparent by writing entries on figures whose lives haven't been completely lost, such as Agnodike and Aglaonike, two ancient Greek women, one a brave physician, the other a beguiling astronomer. And fortunately, information about other remarkable women of science has survived too, thanks in part to pop culture.

Although it wasn't a big hit, *Agora,* a 2009 film, spotlighted an important female astronomer and mathematician in late-4th-century CE Roman Egypt: Hypatia (portrayed by Rachel Weisz). Hypatia's written work was lost in the Library of Alexandria's destruction, but "all our sources agree," says Maria Dzielska, a scholar of the Roman Empire, "that she was a model of ethical courage, righteousness, veracity, civic devotion, and intellectual prowess." Due to her brilliance, her father, Theon, raised Hypatia as Greeks would traditionally raise a son—he taught her his craft, mathematics, and eventually she became the head of a Neo-Platonist school in Alexandria, something only men had previously done. Before she was brutally murdered by a Christian brotherhood, she built medical and astronomical devices as well as an apparatus for distilling water.

Though Hypatia was in many ways an exemplary female figure of science and philosophy, she wasn't a singular figure. Women made strides in the major fields of ancient science.

Medicine and Chemistry

The first recorded woman physician, who was possibly the first woman scientist, was Merit Ptah, an Egyptian living in the 28th century BCE. She was the chief physician of the pharaoh's court during the Second Dynasty, a time when Egyptian women regularly became physicians and midwives, studying at both coed and all-women medical schools. Centuries later, during the Fourth Dynasty (26th–24th century BCE), Peseshet, the administrator of Sais, one such medical school for women, oversaw all women physicians in the empire.

In 13th-century BCE Babylon, intelligent women in the cradle of civilization were able to hold positions of authority. While overseeing the royal palace, a perfumer named Tapputi invented the still, used for purifying substances like alcohol, and perhaps became the world's first chemist.

Artemisia of Caria II, another woman of science, is mainly remembered for her husband, Mausolus (a recurring theme). She's known today for ordering the construction of the mausoleum at Halicarnassus, built between 353 and 351 BCE, as a memorial to her husband (and brother), who she loved so much that she is said to have, on occasion, mixed his ashes in with her drinks. But she was also a botanist and medical researcher. She discovered the myriad uses of the genus *Artemisia* (wormwoods), which is named after her. The herb can stimulate pelvic and uterine blood flow, induce abortion, expel retained placentas, and prevent miscarriage. (In a happy connection, Tu Youyou was co-awarded the 2015 Nobel Prize in Medicine for her discovery of a malaria treatment, artemisinin, derived from one species of wormwood.)

Probably the most famous ancient Greek woman physician was Agnodike, a possible contemporary of Artemisia's. In the 4th century BCE she pretended to be a man to study with Herophilus, the first anatomist. She was motivated to study because of women who died unnecessarily in childbirth, or of reproductive diseases because they did not want to see a male physician. Apparently out of jealousy, the men of Athens decried her because they thought she was seducing women, but they discovered her true identity, and she was put on trial for practicing medicine illegally. Agnodike could have been sentenced to death were it not for the women of Athens, who protested at her trial. As a result a new law was passed allowing women to be physicians.

Metrodora, who lived sometime during the 3rd to 5th century CE, became the first woman medical scholar by authoring a treatise on, among other topics, gynecology, called *On the Diseases and Cures of Women*. It contained herbal remedies found nowhere else in ancient Greek writing, discussed the causes of various types of vaginal discharge, and went on to be widely referenced by succeeding practitioners in ancient Greece and Rome.

Astronomy and Mathematics

Egypt wasn't the only place in antiquity where women flourished in the sciences. In the 23rd century BCE, Akkadian astronomer-priestess En Hedu-Anna perhaps became the world's first known woman astronomer, since her position required her to make detailed astronomical calculations and observations—no small feat, given that writing was only a few centuries old. She served Inanna, goddess of the moon, and is best known for her sweeping sacred poetry that survives to this day. After her death she was elevated to demigoddess status and worshipped by the local Sumerian citizenry. Last year a crater on Mercury was named after her.

Another woman undeservedly overshadowed by her famous male paramour was Theano, typically remembered as Pythagoras's wife. She was an accomplished astronomer and mathematician who led the Pythagoreans after her husband died. Like Theano, Aglaonike—a 1st- or 2nd-century CE Greek astronomer—was intellectually gifted. It was thought that since she could predict the times of lunar eclipses, she was capable of making the moon disappear on demand. She was not shy about her abilities of observation and calculation and was vilified for bragging. The women astronomers who associated themselves with Aglaonike during her life and after her death were called the "witches of Thessaly" because, as Plutarch wrote, Aglaonike "imposed upon the women, and made them all believe that she was drawing down the moon."

Philosophy

By the 5th century BCE ancient Greek civilization was flourishing, and contrary to modern impressions, women were prominent participants in philosophy, mathematics, astronomy, and medicine. Take the philosopher Aspasia of Miletus, a quasi-contemporary of Theano. Her teachings are suggested to have influenced Socrates, and as a hetaera (a high-class courtesan) she advocated for Greek women in society. Although she is mostly remembered today as being Pericles's lover and partner, few recall that it was she who authored his famous funeral oration. (Scholars have remarked on how structurally and thematically similar it is to Lincoln's Gettysburg Address.)

Since hetaerae were meant to be well educated, in order to provide not only sexual companionship but also intellectual stimulation, Aspasia's authorship of one of antiquity's most cherished speeches shouldn't, frankly, be that surprising.

There must have been far more women contributors to the architecture of science and civilization than we know of today. The ravages of time—and humanity—have unfortunately left us with mere scraps of their biographies and treatises, or references to ones lost. If I had a time machine, I'd volunteer in an instant to secure women's vanished intellectual achievements—and I'd bet that the world would be a vastly different place if these ancient women got their due.

Contributors' Notes

Other Notable Science and Nature Writing of 2016

Contributors' Notes

Becca Cudmore is a science journalist from Oregon. She holds an MA from NYU and her work has appeared in *Audubon, The Scientist, Scientific American,* and elsewhere. She will be a Peace Corps volunteer from 2017 to 2019.

Sally Davies is a writer, essayist, and senior editor at *Aeon* magazine. Previously she was the technology and innovation correspondent at the *Financial Times* in London and an associate editor at *Nautilus* magazine in New York. She's currently preoccupied with feminist philosophy, theories of consciousness, and speculative fiction, and lives on Regent's Canal in East London with her partner and their cat.

Robert Draper is a contributing writer for *National Geographic* and a writer at large for *The New York Times Magazine.* He is the author of several books, including the *New York Times* bestseller *Dead Certain: The Presidency of George W. Bush.* He lives in Washington, D.C.

David Epstein is a freelance science writer and investigative reporter and the author of the *New York Times* bestseller *The Sports Gene,* which was chosen by the *Washington Post* as one of the best books of 2013. He was previously an investigative reporter for *ProPublica,* where he worked with editor Tracy Weber on the story in this anthology. Prior to that he was a senior writer at *Sports Illustrated.* He has a master's degree in environmental science.

Sarah Everts is a science journalist who was born in Montreal but is now based in Berlin. She writes regularly about chemistry, art conservation, and the history of science as a European correspondent for *Chemical & Engineering News,* and she freelances on occasion for *Scientific American,*

Smithsonian, Distillations, and *New Scientist* magazines. Sarah is currently working on a book on the science, history, and culture of sweat.

Ann Finkbeiner is a freelance science writer living in Baltimore who writes mostly about astronomy and cosmology. She's written three books and coauthored a fourth, all on wildly different subjects. She's coproprietor of a splendid group science blog (splendid group, splendid blog) called The Last Word on Nothing: www.lastwordonnothing.com.

Azeen Ghorayshi is a science reporter at *BuzzFeed News,* where she reports at the intersection of science and culture. She has previously written for the *Guardian, Newsweek, New Scientist, Wired UK,* and other publications. She has won the AAAS Kavli Science Journalism Award and the Clark/Payne Award for Young Science Journalists and was a finalist in the Livingston Award's national reporting category for her reporting on sexual harassment in science.

Chris Jones wrote for many years at *Esquire,* where he won two National Magazine Awards for his feature writing and wrote often about space. "The Woman Who Might Find Us Another Earth" was his first piece for *The New York Times Magazine* and is his first appearance in *The Best American Science and Nature Writing.*

Kathryn Joyce is a journalist and the author of *The Child Catchers* and *Quiverfull.* Her work has appeared in *Highline, Pacific Standard,* the *New Republic, Mother Jones,* and many other publications.

Tom Kizzia was a longtime reporter for the now-departed *Anchorage Daily News.* He is the author of *Pilgrim's Wilderness: A True Story of Faith and Madness on the Alaska Frontier,* a *New York Times* bestseller and Amazon top-ten book of the year. His first book, *The Wake of the Unseen Object,* explored the changing ways of rural Alaska Native culture in the modern world. He lives in Homer, Alaska.

Elizabeth Kolbert is a staff writer for *The New Yorker* and the author of *The Sixth Extinction,* which won the 2015 Pulitzer Prize for general nonfiction. She is also the author of *Field Notes from a Catastrophe: Man, Nature, and Climate Change,* which grew out of a three-part series titled "The Climate of Man." Kolbert is a two-time National Magazine Award winner and has received a Heinz Award, a Lannan Literary Fellowship, and the Rose-Walters Prize. She lives in Williamstown, Massachusetts.

Maria Konnikova is the author of two *New York Times* bestsellers, *The Confidence Game,* winner of the 2016 Robert P. Balles Prize in Critical Thinking, and *Mastermind: How to Think Like Sherlock Holmes,* an Anthony and Agatha Award finalist. She is a contributing writer for *The New Yorker* and is currently working on a book about poker and the balance of skill and luck in life, to be published in 2019. Maria is also the host of the podcast *The Grift* from Panoply Media, a show that explores con artists and the lives they ruin. She graduated from Harvard University and received her PhD in psychology from Columbia University.

Adrian Glick Kudler grew up in New England but has lived in Los Angeles for a pretty long time. She is the West Coast features editor for *Curbed.*

Jon Mooallem is writer at large at *The New York Times Magazine* and a contributor to *This American Life.* He is the author of *American Hippopotamus* and *Wild Ones: A Sometimes Dismaying, Weirdly Reassuring Story About Looking at People Looking at Animals in America.*

Omar Mouallem is a Canadian National Magazine Award–winning writer whose stories have appeared in *NewYorker.com, Wired,* and *The Guardian.* He coauthored a book about the 2016 Alberta wildfires, titled *Inside the Inferno: A Firefighter's Story of the Brotherhood That Saved Fort McMurray.* He lives in Edmonton, Alberta, and tweets at @omar_aok.

Michelle Nijhuis writes for *National Geographic* and *The Atlantic,* blogs for *The New Yorker,* and is a longtime contributing editor of *High Country News.* She is the coeditor of *The Science Writers' Handbook* and the author of *The Science Writers' Essay Handbook,* and her reporting on conservation and global change has won several national awards. After 15 years off the electrical grid in rural Colorado, she and her family now live in White Salmon, Washington.

Tom Philpott is the food and agriculture correspondent for *Mother Jones.*

Michael Regnier is a science writer and editor at the Wellcome Trust. He has a degree in natural sciences and a master's in science communication, which included a brief stint in Geneva helping to develop public exhibitions at CERN. Before writing about science for a living, he wrote plays, the best of which was *Time Out's* Critic's Choice when it was performed in London in 2002. He still enjoys theater and spent the opening weekend of the 2016 Edinburgh Festival writing "diagnoses" of shows relating to health for a Wellcome-funded project called "The Sick of the Fringe." Michael lives in London with his wife, two daughters, and one cat.

Nathaniel Rich is a writer at large for *The New York Times Magazine* and a regular contributor to *The New York Review of Books* and *The Atlantic*. He is the author of three novels: *The Mayor's Tongue, Odds Against Tomorrow,* and *King Zeno* (2018).

Sonia Smith is an associate editor at *Texas Monthly*. She has also written for *Slate,* the *New York Times, Roads & Kingdoms,* and the *Kyiv Post*. She lives in Dallas with her husband and two dogs.

Christopher Solomon (www.chrissolomon.net) has written for publications ranging from *The New York Times Sunday Magazine* to *Scientific American* to *Food & Wine*. He is a contributing editor at *Outside* and *Runner's World*. His stories have appeared in *The Best American Sports Writing* and twice in *The Best American Travel Writing*. He lives in Seattle.

Emily Temple-Wood is a current student and future doctor. She is a Wikipedia editor, working to correct systemic bias against women through WikiProject Women Scientists. Emily lives in the suburbs of Chicago and owns far too many books.

Kim Tingley is a contributing writer for *The New York Times Magazine*. She has been recognized with a Rona Jaffe Foundation Writers' Award and a fellowship from the Nieman Foundation for Journalism at Harvard. She holds an MFA from Columbia University.

Nicola Twilley is a cohost of the award-winning *Gastropod* podcast, author of the blog *Edible Geography,* and a contributing writer at *The New Yorker*. She is deeply obsessed with refrigeration and is currently writing a book on the topic. She is also coauthoring a book about the past, present, and future of quarantine with Geoff Manaugh. In her spare time she makes smog meringues as part of an ongoing exploration of the taste of "aeroir" with the Center for Genomic Gastronomy. "The Billion-Year Wave" was edited by Anthony Lydgate.

Other Notable Science and Nature Writing of 2016

SELECTED BY TIM FOLGER

THE BEST AMERICAN SERIES®

FIRST, BEST, AND BEST-SELLING

The Best American Comics

The Best American Essays

The Best American Mystery Stories

The Best American Nonrequired Reading

The Best American Science and Nature Writing

The Best American Science Fiction and Fantasy

The Best American Short Stories

The Best American Sports Writing

The Best American Travel Writing

Available in print and e-book wherever books are sold.
Visit our website: *www.hmhco.com/bestamerican*